Increasing Fatigue Strength of Austenitic Stainless Steel Using Machine Hammer Peening

Erhöhung der Ermüdungsfestigkeit von austenitischem rostfreien Edelstahl durch maschinelles Oberflächenhämmern

Von der Fakultät für Maschinenwesen
der Rheinisch-Westfälischen Technischen Hochschule Aachen
zur Erlangung des akademischen Grades eines
Doktors der Ingenieurwissenschaften
genehmigte Dissertation

vorgelegt von

Robby Mannens

Berichter/in:

Univ.-Prof. Dr.-Ing. Thomas Bergs
Univ.-Prof. Dr.-Ing. Sebastian Münstermann

Tag der mündlichen Prüfung: 20. August 2021

ERGEBNISSE AUS DER PRODUKTIONSTECHNIK

Robby Mannens

Increasing Fatigue Strength of Austenitic Stainless Steel Using Machine Hammer Peening

Herausgeber:
Prof. Dr.-Ing. T. Bergs
Prof. Dr.-Ing. Dipl.-Wirt. Ing. G. Schuh
Prof. Dr.-Ing. C. Brecher
Prof. Dr.-Ing. R. H. Schmitt

Band 25/2021

Bibliografische Information der Deutschen Nationalbibliothek
Die Deutsche Nationalbibliothek verzeichnet diese Publikation in der Deutschen Nationalbibliografie; detaillierte bibliografische Daten sind im Internet über https://portal.dnb.de abrufbar.

Robby Mannens:

Increasing Fatigue Strength of Austenitic Stainless Steel Using Machine Hammer Peening

1. Auflage, 2021

Gedruckt auf holz- und säurefreiem Papier, 100% chlorfrei gebleicht.

Apprimus Verlag, Aachen, 2021
Wissenschaftsverlag des Instituts für Industriekommunikation und Fachmedien an der RWTH Aachen
Steinbachstr. 25, 52074 Aachen
Internet: www.apprimus-verlag.de, E-Mail: info@apprimus-verlag.de

Printed in Germany

ISBN 978-3-98555-006-7

D 82 (Diss. RWTH Aachen University, 2021)

Vorwort und Danksagung

Preamble and Acknowledgement

"Much to learn you still have.
This is just the beginning."

YODA, Jedi Master

Die vorliegende Arbeit entstand während meiner Tätigkeit als wissenschaftlicher Mitarbeiter am Werkzeugmaschinenlabor (WZL) der RWTH Aachen University. Mein besonderer Dank gilt Herrn Univ.-Prof. Dr.-Ing. Thomas Bergs, Inhaber des Lehrstuhls für Technologie der Fertigungsverfahren, und Univ.-Prof. Dr.-Ing. Dr.-Ing. E.h. Dr. h.c. Dr. h.c. Fritz Klocke, ehemaliger Inhaber des Lehrstuhls für Technologie der Fertigungsverfahren, für die fachliche und persönliche Förderung sowie die stetige Unterstützung meiner Arbeiten.

Weiterhin gilt mein Dank Herrn Univ.-Prof. Dr.-Ing. Sebastian Münstermann, Leiter des Lehr- und Forschungsgebietes für Werkstoff- und Bauteilintegrität am Institut für Eisenhüttenkunde (IEHK) der RWTH Aachen University, für die eingehende Durchsicht des Manuskriptes und die Übernahme des Koreferats. Herrn Univ.-Prof. Dr. rer. nat. Constantin Häfner, Inhaber des Lehrstuhls für Lasertechnik (LLT) an der RWTH Aachen University, danke ich für die Übernahme des Prüfungsvorsitzes.

Die Deutsche Forschungsgemeinschaft (DFG) hat einen großen Teil der zugrundeliegenden Forschungsarbeiten im Rahmen des Forschungsprojektes *Kornfeinung in der Randzone von metallischen Werkstoffen durch maschinelles Oberflächenhämmern (FinePeening)* gefördert. Wichtige Inhalte wurden weiterhin im Rahmen des vom Bundesministeriums für Wirtschaft und Energie (BMWi) geförderten Mittelstand 4.0-Kompetenzzentrums durch das Projekt *Vernetztes maschinelles Oberflächenhämmern* erarbeitet. Hierfür möchte ich mich herzlich bei der DFG, dem BMWi sowie ihren Mitarbeitern bedanken. Den Projektpartnern um Christian Löcker (accurapuls GmbH) und Dr. rer. nat. Alexander Schwedt (Gemeinschaftslabor für Elektronenmikroskopie der RWTH Aachen University, GfE) gilt mein besonderer Dank.

Ebenso gilt mein Dank den Kollegen des Workshop Machine Hammer Peening, besonders Peter Sticht (Prof. P. Groche, PtU, TU Darmstadt), Dr.-Ing. Alexander Klumpp (Prof. V. Schulze, wbk, KIT) sowie Dr.techn. Stephan Krall und Dr.techn. Christoph Lechner (Prof. F. Bleicher, IFT, TU Wien) für die inspirierenden und unterstützenden Diskussionen.

Besonders bedanken möchte ich mich bei meinen aktuellen und ehemaligen Kollegen und Freunden am Werkzeugmaschinenlabor WZL der RWTH Aachen für die wertvolle Unterstützung und die freundschaftliche Zusammenarbeit in den letzten Jahren. Hervorheben möchte ich an dieser Stelle Herr Dr.-Ing. Dipl.-Wirt.Ing. Daniel Trauth und Herr Dr.-Ing. Andreas Feuerhack. Es war ein Privileg für mich, von euch lernen zu dürfen und mich dank eurer Fürsprache und Unterstützung sowohl fachlich als auch persönlich weiterentwickeln zu können. Dr.-Ing. Patrick Mattfeld danke ich herzlichst für das Vertrauen und die Möglichkeit einer Promotion im Bereich Umformtechnik am WZL. Meinen Freunden Dr.-Ing. Maximilian Lachenmaier und Dr.-Ing. Jan Rey danke ich für den fortwährenden Austausch und die anregenden Gespräche in unserer Gruppenleiterzeit während, aber vor allem auch außerhalb der Arbeitszeit. Ebenso möchte ich mich ganz besonders bei Dr.-Ing. Rafael Hild und Dr.-Ing. Herman Voigts für eure Unterstützung, Freundschaft und die außergewöhnliche Zeit auf der Arbeit sowie abseits davon in entspannter Atmosphäre bedanken. Meinen Umformerkollegen Dr.-Ing. Anton Shirobokov, Dr.-Ing. Oksana Baer, Dr.-Ing. Tim Herrig, Frank Schweinshaupt, Lars Uhlmann, Ingo Felix Weiser, Matthias Nick, Martina Müller, Thomas Kokott und Marcel Botz gebührt ebenfalls mein Dank für die fortwährende und umfangreiche Unterstützung sowie die tollen gemeinsamen Momente.

Danke sagen möchte ich auch meinen zahlreichen aktiven und ehemaligen Kollegen und Freunden der Abteilungen Technologieplanung und Schleiftechnik, Abtragende Fertigungsverfahren, Digitale Transformation, Zerspantechnologie und Getriebetechnik, mit denen ich vor allem im Rahmen gemeinsamer Kegelabende und Skitouren auch abseits fachlicher Themen eine tolle Zeit verbringen durfte. Den Mitarbeitern des WZL in der mechanischen Werkstatt, der Dokumentation, der EDV, der Metallografie und der Messtechnik danke ich für ihre fortwährende Unterstützung meiner Arbeiten.

Meinen studentischen Mitarbeitern sowie Studien-, Bachelor- und Masterarbeitern danke ich für ihren tollen Einsatz und ihren großen Anteil an der Erstellung dieser Arbeit. Ganz besonders möchte ich mich bei Felix Lambers, Felix Metz sowie Rainer Radloff und Peter Herrmann (beide mittlerweile Umformerkollegen) bedanken. Ihr habt mich sehr entlastet. Vielen Dank!

Meine ältesten und engsten Freunde aus meiner belgischen Heimat waren in vielen guten wie auch in schwierigen Zeiten stets eine Stütze für mich. Für eure tiefe Freundschaft bin ich mehr als dankbar.

Mein größter Dank gilt meiner Familie, auf deren Unterstützung und Rückhalt ich mich immer verlassen konnte. Insbesondere möchte ich mich bei meinen Eltern Elly Renardy, Raymond Mannens und Guido Lausberg bedanken. Durch eure bedingungslose Unterstützung und euren Glauben an mich habt ihr zum erfolgreichen Abschluss dieser Promotion maßgeblich beigetragen.

Für ihre liebevolle Unterstützung auch in schwierigen Zeiten, ihren Glauben an mich sowie ihr Verständnis für die langen Arbeitstage und die kurzen Wochenenden danke ich von ganzem Herzen meiner Freundin Alice. Unsere hervorragende Teamarbeit war ausschlaggebend für meine erfolgreiche Promotion. Sie wäre mir ohne deinen Rückhalt nicht gelungen.

Aachen, September 2021 Robby Mannens

Abstract

One way to reduce CO_2 emissions is to develop renewable energies. Energy conversion plants contain many mechanical components made of stainless steel, which have to withstand complex load collectives. However, the performance of such components is limited due to inadequate mechanical strength and fatigue properties, especially in the highly loaded surface layer.

One way to improve these properties is to use forming surface layer treatments (SLT), such as electrodynamic machine hammer peening (MHP). MHP is a cold micro forging process that machines metallic surface layers at high frequency using a spherical plunger. The MHP tool is usually guided by machining centers. Due to the innovative character of robot-guided MHP, there are currently still fundamental knowledge deficits in the area of process kinetics, contact mechanisms and the surface layer and fatigue properties of hammered materials. MHP machining of stainless steels is still largely unexplored.

The subject of this work is, in addition to the development of a knowledge base on MHP process kinetics and the influence of industrial robots as MHP carrier systems, the analysis of contact mechanisms and surface layer and fatigue properties of metastable austenitic stainless steel. In a first step, an electrodynamic MHP system was modified with inline force and distance sensors, characterized, and a databased numerical MHP kinetic model was developed using machine learning methods. In a second step, single-impact phenomena were investigated experimentally by means of a drop tower test rig with regard to geometric and microstructural mechanisms as a function of the energy input and supplemented by numerical FE simulations of the plastomechanical material behavior. In a third step, cause-effect relationships between the energy input and the material-physical mechanisms were determined by means of robot-based MHP tests and synthesized with findings from dislocation-dynamic FE simulations to form a design model. In the final step, the fatigue behavior of hammered workpieces was evaluated in rotating bending tests in accordance with the developed model.

Kurzzusammenfassung

Eine Möglichkeit die globalen CO_2-Emissionen zu reduzieren, besteht im Ausbau erneuerbarer Energien. Die Anlagen zur Energieumwandlung beinhalten viele mechanische Komponenten aus rostfreiem Edelstahl, welche komplexen Belastungskollektiven standhalten müssen. Die Leistungsfähigkeit solcher Komponenten ist jedoch aufgrund der unzureichenden mechanischen Festigkeits- und Ermüdungseigenschaften, besonders im hochbelasteten Randzonenbereich, begrenzt.

Eine Möglichkeit diese Eigenschaften zu verbessern, besteht im Einsatz von umformenden Randzonenbearbeitungsverfahren, wie dem elektrodynamischen maschinellen Oberflächenhämmern (MOH). Das MOH ist ein Kaltmikroschmiedeverfahren, welches metallische Randzonen hochfrequent mittels eines sphärischen Stößels bearbeitet. Die Führung des MOH-Werkzeugs erfolgt dabei zumeist über Bearbeitungszentren. Aufgrund des innovativen Charakters des robotergeführten MOH existieren aktuell noch grundlegende Wissensdefizite im Bereich der Prozesskinetik, der Kontaktmechanismen sowie der Randzonen- und Ermüdungseigenschaften gehämmerter Werkstoffe. Die MOH-Bearbeitung von rostfreien Edelstählen ist hierbei noch größtenteils unerforscht.

Gegenstand dieser Arbeit ist neben dem Aufbau einer Wissensbasis zur MOH-Prozesskinetik sowie zum Einfluss von Industrierobotern als MOH-Trägersystem auch die Analyse der Kontaktmechanismen und der Randzonen- und Ermüdungseigenschaften von metastabilem austenitischen Edelstahl. In einem ersten Schritt wurde ein elektrodynamisches MOH-System mit Inline-Kraft- und Abstandssensorik aufgerüstet, charakterisiert und mittels Methoden des maschinellen Lernens ein datenbasiertes, numerisches MOH-Kinetikmodell entwickelt. In einem zweiten Schritt wurden experimentell mittels eines Fallturmprüfstandes Einzeleinschlagsphänomene hinsichtlich geometrischer und mikrostruktureller Mechanismen in Abhängigkeit des Energieeintrags untersucht und durch numerische FE-Simulationen zum plastomechanischen Werkstoffverhalten ergänzt. In einem dritten Schritt wurden mittels roboterbasierter MOH-Versuche Ursache-Wirkungzusammenhänge zwischen dem Energieeintrag sowie den werkstoffphysikalischen Mechanismen ermittelt und mit Erkenntnissen aus versetzungsdynamischen FE-Simulationen zu einem Gestaltungsmodell synthetisiert. Im letzten Schritt wurden entsprechend des erarbeiteten Modells gehämmerte Werkstücke in Umlaufbiegeversuchen hinsichtlich des Ermüdungsverhalten bewertet.

Table of contents

Inhaltsverzeichnis

1 Introduction ... 1

2 State of the art in research and industry ... 9

2.1 Austenitic stainless steels ... 9
2.1.1 Austenite stability ... 10
2.1.2 Martensitic transformation ... 12
2.1.3 Deformation behavior of austenitic stainless steels ... 17
2.1.4 Intermediate conclusion ... 19
2.2 Surface layer treatment ... 20
2.2.1 Processes for mechanical surface layer improvement ... 20
2.2.2 Effects on surface layer ... 21
2.2.3 Influence on fatigue behavior ... 26
2.2.4 Intermediate conclusion ... 29
2.3 Machine hammer peening (MHP) ... 29
2.3.1 Technological process variations ... 30
2.3.2 Electrodynamic operating principle ... 31
2.3.3 Kinematics, contact mechanics and impact analysis ... 32
2.3.4 Intermediate conclusion ... 35
2.4 Economic and scientific problem statement ... 35

3 Objective and research methodology ... 37

4 Characterization and analysis of robot-based MHP kinetics ... 41

4.1 Determination of robot dynamics ... 42
4.1.1 Analysis of robot acceleration ... 42
4.1.2 Examination of robot oscillations ... 44
4.1.3 Influence of robot dynamics on MHP processing ... 47
4.2 Metrological extension of the electrodynamic MHP system ... 48
4.2.1 Requirement analysis and sensor technology selection ... 48
4.2.2 Instrumentation of MHP tool ... 49
4.2.3 Influence of sampling rate on process data ... 51
4.3 Effects of process parameters on MHP system behavior ... 53
4.3.1 Test setup and experimental procedure ... 53
4.3.2 Free stroke analysis ... 53
4.3.3 Determination of current, voltage and electrical power ... 55
4.3.4 Determination of hammering forces ... 56
4.3.5 Determination of plunger stroke and contact energy ... 57
4.3.6 Deduction of an empirical-mathematical MHP model ... 60
4.4 Numerical modelling of multiphysical MHP system behavior ... 61
4.4.1 Electrical operating system ... 62
4.4.2 Mechanical operating system ... 65

4.4.3 Contact mechanics ... 66
4.4.4 Deduction of a data-based MHP kinetics model ... 67
4.5 Intermediate conclusion ... 70

5 Plastomechanical single impact analysis ... 73

5.1 Development of drop tower test rig ... 73
5.1.1 Test rig design for single impact analysis ... 74
5.1.2 Evaluation of result transferability to MHP process ... 75
5.2 Experimental analysis of indentation geometry and contact modes ... 76
5.2.1 Procedure and evaluation of indentations ... 77
5.2.2 Results of contact conditions and indentation morphology ... 78
5.3 Numerical FE modeling of impact tests ... 84
5.3.1 Set-up and verification of the FE model ... 84
5.3.2 Analysis of plastomechanical material state ... 86
5.4 Intermediate conclusion ... 90

6 Influence of MHP on surface integrity ... 93

6.1 Experimental analysis of surface layer after robot-based MHP ... 93
6.1.1 MHP test setup and procedure ... 93
6.1.2 Analysis methods and specimen preparation ... 95
6.1.3 Results of surface layer characterization ... 96
6.2 Numerical FE modeling of machine hammer peening ... 109
6.2.1 Set-up and validation of the FE model ... 110
6.2.2 Analysis of stress states and dislocation density ... 114
6.3 Intermediate conclusion ... 118
6.4 Synthesis of results into design model ... 119

7 Fatigue analysis of MHP-treated surface layers ... 123

7.1 Experimental design and procedure ... 123
7.1.1 Extension of MHP test bench and test conditions ... 123
7.1.2 Validation of rotating bending test rig and approach ... 126
7.2 Analysis of load cycles and fractographs ... 128
7.3 Conclusion and validation of the research hypothesis ... 132

8 Summary and outlook ... 133

Bibliography ... XIII

Appendix ... XXXVII

Symbols and abbreviations

Formelzeichen und Abkürzungsverzeichnis

Capital letters

A	%	Elongation at break
A_{i}	mm²	Projected area of indentation
A_{JC}	N·mm⁻²	JOHNSON-COOK model coefficient
A_{L}	-	Sum of the products of f_{i} and i
A_{PL}	-	Coefficient of power law
B_{JC}	N·mm⁻²	JOHNSON-COOK model coefficient
B_{l}	Tm	Magnetic field action coefficient
B_{L}	-	Numerical constant in LEMIALE material model
B_{PL}	-	Coefficient of power law
C_{JC}	-	JOHNSON-COOK model coefficient
Cr_{eq}	%	Chrome equivalent
E	N·mm⁻²	YOUNG'S modulus
E_{c}	mJ	Contact energy
$E_{\mathrm{c,max}}$	mJ	Maximum contact energy
E_{E}	Wh	Electricity generation
E_{el}	mJ	Elastic energy at the workpiece
E_{Fr}	mJ	Frictional energy
E_{kin}	mJ	Kinetic energy
$E_{\mathrm{kin,E}}$	mJ	Kinetic energy at the end of the forming step
$E_{\mathrm{kin,S}}$	mJ	Kinetic energy at the beginning of the forming step
E_{pl}	mJ	Plastic energy at the workpiece
E_{pot}	mJ	Potential energy
F	-	Sum of the events f_{i} on all evaluable load steps
F_{G}	N	Gravitational force
F_{h}	kN	Hammering force in MHP
$F_{\mathrm{h,max}}$	kN	Maximum hammering force with new MHP system
$F_{\mathrm{h,max,old}}$	kN	Maximum hammering force with old MHP system

F_{i}	N	Indentation force in micro hardness measurement
F_{I}	N	Contact force at inner stop in MHP tool
F_{L}	kN	LORENTZ force
F_{Lathe}	kN	Lathe force
F_{prep}	N	Polishing force in preparation of microstructure
F_{r}	N	Friction force
F_{reac}	kN	Reaction force
F_{WP}	N	Contact force at workpiece surface
G	J	Free GIBBS energy
G_{L}	$\mathrm{N{\cdot}mm^{-2}}$	Shear modulus
$HV0.1$	$\mathrm{N{\cdot}mm^{-2}}$	Vickers micro hardness for a test force F_{t} = 0.9807 N
$HV5$	$\mathrm{N{\cdot}mm^{-2}}$	Vickers macro hardness for a test force F_{t} = 49.03 N
I	A	Input current in MHP
$I_{1...3}$	N·s	Impulse of falling mass
K	-	Numerical constant in LEMIALE material model
L	mH	Inductance in MHP
M	-	TAYLOR factor
M_{b}	Nm	Bending moment
M_{s}	°C	Martensite starting temperature
M_{d}	°C	Martensite deformation temperature
M_{d30}	°C	Temperature where 50% martensite has formed after 30% true strain
N_{L}	-	Number of load cycles
Ni_{eq}	%	Nickel equivalent
P	W	Input electrical power in MHP
P_{el}	%	Electrical power stage in MHP
P_{x}	$\mathrm{W{\cdot}m^{-2}}$	Radiative forcing
Ra	µm	Arithmetical mean roughness
R_{m}	$\mathrm{N{\cdot}mm^{-2}}$	Tensile strength
R_{mhp}	Ω	Electrical resistance in MHP
$R_{\mathrm{p0.2}}$	$\mathrm{N{\cdot}mm^{-2}}$	Yield strength

Sa	µm	Average mean roughness
Sz	µm	Root mean square roughness
T	°C	Temperature
T_0	°C	Thermodynamic equilibrium temperature
U, $U_0(t)$	V	Input voltage in MHP
U_a	V	Accelerating voltage in EBSD analysis
U_i	V	Resulting second voltage acting contrary to U_0
V_i	µm³	Indentation volume
V_p	t	World stainless steel production
V_pl	mm³	Volume of plastically deformed zone of indentation
$V_\mathrm{SS,China}$	USD	Stainless steel market size in China
Z	%	Elongation

Lower case letters

a	mm	Indentation distance in MHP
a_p	mm	Infeed in grinding
a_r	m·s⁻²	Acceleration of robot
$a_{1...5}$	-	Numerical values of experimental-mathematical model
b	-	BASQUIN coefficient
b_B	m	BURGERS vector
c	-	COFFIN-MANSON coefficient
c_ax	N·µm⁻¹	Axial stiffness of vibrating MHP mass
c_I	-	Damping parameter at inner stop in MHP tool
$c_{1...10}$	-	Effect parameters in HV5 description model
d	mm	Hammer head diameter
d_c	µm	Average dislocation cell size
$\bar{d}_\mathrm{G}$	µm	Mean grain size
d_gc	µm	Grain / cell size
d_i	mm	Indentation diameter
d_L	-	Mean logarithmic step change
d_m	µm	Diameter of structured area on MHP plunger

d_s	mm	Diameter of anodized shaft in drop tower test rig
d_γ	µm	Austenite grain size
d_w	mm	Working distance in EBSD analysis
d_σ/d_ε	$N \cdot mm^{-2}$	Work hardening rate
e	-	Coefficient of restitution
e_I	-	Coefficient of restitution at the inner stop in MHP tool
e_c	-	Coefficient of restitution at the workpiece
f	Hz	Hammering frequency
f_a, f_b	-	Damping parameters at the workpiece surface
f_i	-	Events on one evaluable load step
f_{RB}	Hz	Frequency in rotating bending test
f_s	Hz	Sampling frequency
f_w	-	Volume fraction of dislocation walls
f_0	-	Volume fraction of dislocation walls in initial state
f_∞	-	Volume fraction of dislocation walls in saturated state
g	$m \cdot s^{-2}$	Gravitational acceleration
h	mm	Plunger stroke in MHP
h_d	mm	Drop height in drop tower test rig
h_{dt}	mm	Total height of drop tower test rig
h_{max}	mm	Maximum plunger stroke in MHP
h_{min}	mm	Minimum plunger stroke in MHP
h_{RC20}	mm	Measuring range of fiber optic distance sensor
$h_{WP,FE}$	mm	Height of cuboidal workpiece in FEM
i	-	Load level number
$i_{\alpha'}$	-	Fraction of the α'-martensite phase
i_γ	-	Fraction of the γ-austenite phase
i_{Twins}	-	Fraction of twins in the austenite phase
k_0	-	Numerical parameter in LEMIALE material model
l_e	µm	Element size in FE simulation
l_L	mm	Lever arm
l_{ra}	m	Working range of industrial robot

$l_{WP,FE}$	mm	Length of cuboidal workpiece in FEM
l_x	mm	Path length in MHP
m	g	Mass of oscillating plunger
$\dot{m}$	$m^3 \cdot s^{-1}$	Mass flow rate
m_{B5000}	g	Mass of piezo force sensor KISTLER 9712B5000
m_{CO_2}	t	CO_2 emissions
m_{er}	g	Mass of epoxy resin with stainless steel plate in MHP
m_{FP}	g	Mass of force measuring plunger in MHP
m_L	-	Strain rate sensitivity parameter in LEMIALE model
m_{ob}	g	Mass of oscillating body in MHP tool
$m_{Oil,eq.}$	t	World primary energy demand
m_p	g	Mass of MHP plunger
m_{p6}	g	Mass of plunger with 6 mm hammer head diameter
m_{p12}	g	Mass of plunger with 12 mm hammer head diameter
m_r	kg	Handling capacity of industrial robot
m_{tot}	kg	Total mass of MHP tool
n	-	Number of overruns in MHP
n_{EBSD}	nm	Step size in EBSD analysis
n_{JC}	-	JOHNSON-COOK model coefficient
n_L	-	Exponent in LEMIALE material model
n_s	$m \cdot s^{-1}$	Rate of stroke
n_{SB}	-	Number of shear band intersections
o_i	%	Overlap of indentation
$r_{a...d}$	-	Friction parameters in multiphysical MHP kinetic model
s	mm	Stepover distance in MHP
t_{bp}	mm	Thickness of base plate in drop tower test rig
t_c	ms	Contact time of plunger with workpiece in MHP
t_{mhp}	s	Manufacturing time in MHP
u_{res}	µm	Deflection of rotating bending specimen
$u_{r,max}$	µm	Maximum elastic compression of vibrating MHP mass
v	$mm \cdot s^{-1}$	Machine feed rate

v_{c}	mm·s^{-1}	Impact velocity
v_{cu}	m·s^{-1}	Cutting speed in grinding
$v_{\mathrm{c,max}}$	mm·s^{-1}	Maximum impact velocity
$v_{\mathrm{p}}(x,t)$	mm·s^{-1}	Time and location dependent plunger velocity
v_{r}	mm·s^{-1}	Robot feed rate
w	J·mm^{-2}	Energy density in MHP
$w_{\mathrm{WP,FE}}$	mm	Width of cuboidal workpiece in FEM
x_{di}	-	Position of one indentation along the MHP feed axis
x_{I}	-	Plunger position relative to the MHP tool
x_{WP}	-	Plunger position relative to the workpiece
z	mm	Surface layer depth
z_{i}	µm	Depth of indentation
z_{m}	µm	Depth of structured area on MHP plunger
z_{pl}	mm	Depth of plastically deformed zone under indentation
$z_{\sigma_{\mathrm{rs,max}}}$	mm	Depth of maximum compressive residual stresses

Greek letters

α	-	α-ferrite phase
α_{L}	-	Numerical constant in LEMIALE material model
α_{mhp}	°	Path orientation angle in MHP
α'	-	α'-martensite phase
α^*	-	Numerical constant in LEMIALE material model
β_{a}	°	Angle of attack in MHP
β_{i}	°	Resulting impact angle in MHP
β_{t}	°	Tilt angle in MHP
β^*	-	Numerical constant in LEMIALE material model
γ	-	Austenite phase
γ^{r}	-	Resolved shear strain
$\tilde{\gamma}^{\mathrm{r}}$	-	Variation of volume fraction of dislocation walls
$\dot{\gamma}^{\mathrm{r}}$	s^{-1}	Shear rate
$\dot{\gamma}^{0}$	s^{-1}	Reference shear rate

γ_{SF}	mJ·m[2]	Stacking fault energy
$\dot{\gamma}_{0,\mathrm{s}}$	s[-1]	Reference shear rate
δ	-	δ-ferrite phase
$\Delta\sigma_{\mathrm{B}}$	%	Bending stress steps
ΔE_{c}	%	Deviations of contact energies
ΔF_{ax}	kN	Measuring range in axial hammering direction
ΔF_{h}	%	Maximum deviation of measured force peaks in MHP
ΔG	J	Free GIBBS energy difference
ΔG_{s}	J	Free GIBBS energy difference at room temperature
ΔG_{cf}	J	Free GIBBS energy difference by cold forming
Δh	µm	Determination accuracy of plunger stroke
Δh_{RC20}	µm	Resolution of fiber optic distance sensor
$\Delta HV5$	%	Deviation in macro hardness
Δt	s	Time delay of plunger contact with workpiece
Δv_{c}	%	Deviations of impact velocities
Δy_{max}	µm	Maximum lateral deviation of indentations in MHP
Δz_{h}	µm	Maximum movement of MHP hammer tool
ε	-	ε-martensite phase
$\varepsilon_{\mathrm{a,el}}$	-	Elastic component of total engineering strain
$\varepsilon_{\mathrm{a,pl}}$	-	Plastic component of total engineering strain
$\varepsilon_{\mathrm{a,tot}}$	-	Total engineering strain
$\varepsilon'_{\mathrm{f}}$	N·mm[-2]	Cyclic ductility coefficient (COFFIN-MANSON coefficient)
$\varepsilon_{\mathrm{pl}}$	-	True plastic strain
$\varepsilon_{\mathrm{pl,max}}$	-	Maximum true plastic strain
$\varepsilon_{\mathrm{tot}}$	-	True total strain
$\dot{\varepsilon}_{\mathrm{pl}}$	s[-1]	Plastic strain rate
$\dot{\varepsilon}_{\mathrm{tot}}$	s[-1]	Total strain rate
ϑ_{M}	-	Volume fraction martensite
μ	-	Coefficient of friction
ν	-	POISSON'S ratio
ρ	kg·mm[3]	Material density

ρ_c	m^{-2}	Cell dislocation density
ρ_{tot}	m^{-2}	Total dislocation density
$\rho_{tot,max}$	m^{-2}	Maximum total dislocation density
$\rho_{tot,0}$	m^{-2}	Total dislocation density in initial state
ρ_w	m^{-2}	Wall dislocation density
$\dot{\rho}_c$	$m^{-2} \cdot s^{-1}$	Evolution of cell dislocation density
$\dot{\rho}_w$	$m^{-2} \cdot s^{-1}$	Evolution of wall dislocation density
σ	$N \cdot mm^{-2}$	Stresses
$\sigma_{b,0}$	$N \cdot mm^{-2}$	Initial bending stress
$\sigma_{b,L}$, $\sigma_{b,crit}$	$N \cdot mm^{-2}$	Critical bending stress
σ'_f	$N \cdot mm^{-2}$	Cyclic fatigue strength coefficient (BASQUIN coefficient)
σ_m	$N \cdot mm^{-2}$	Constant cyclic mean stress
$\sigma_{\|\|}$	$N \cdot mm^{-2}$	Residual stresses parallel to a direction
$\sigma_{\perp}$	$N \cdot mm^{-2}$	Residual stresses perpendicular to a direction
$\sigma_{rs,max}$	$N \cdot mm^{-2}$	Maximum compressive residual stresses
σ_t	$N \cdot mm^{-2}$	Stress amplitude
σ_{vM}	$N \cdot mm^{-2}$	Equivalent (VON MISES) stress
σ_1	$N \cdot mm^{-2}$	Strain rate independent component of σ_{vM}
σ_2	$N \cdot mm^{-2}$	Strain rate dependent component of σ_{vM}
τ^r	$N \cdot mm^{-2}$	Resolved shear stress
τ_c^r	$N \cdot mm^{-2}$	Resolved shear stress of the dislocation cell interior
τ_w^r	$N \cdot mm^{-2}$	Resolved shear stress of the dislocation walls

Acronyms

3D	Three-dimensional
AGGI	Annual Greenhouse Gas Index
AISI	AMERICAN IRON AND STEEL INSTITUT
AW	Aluminum Wrought
Bcc	Body-centered cubic
C3D4	Four-node tetrahedral element (ABAQUS)
C3D8R	Eight-node brick element with reduced integration (ABAQUS)

CDD	Continuum dislocation dynamics
CH_4	Methane
CNC	Computerized numerical control
CO_2	Carbon dioxide
DC	Direct current
DDD	Discrete dislocation dynamics
DIN	DEUTSCHES INSTITUT FÜR NORMUNG
DOE	Design of Experiments
DR	Deep rolling
EBSD	Electron backscatter diffraction
EDX	Energy dispersive X-ray
EN	European Standard
Fcc	Face-centered cubic
FEM	Finite element method
GmbH	Gesellschaft mit beschränkter Haftung (company with lim. liability)
GOM	Gesellschaft für optische Messtechnik
HCF	High cycle fatigue
Hcp	Hexagonal close-packed
HRC	Hardness type C acc. to ROCKWELL
HV	Hardness type acc. to VICKERS
IBF	INSTITUT FÜR BILDSAME FORMGEBUNG (Insitut of Metal Forming)
IEA	INTERNATIONAL ENERGY AGENCY
IEPE	Integrated electronics piezo-electric
IPF	Inverse pole figure
ISO	INTERNATIONAL ORGANIZATION FOR STANDARDIZATION
KAM	Kernel average misorientation
LCF	Low cycle fatigue
LM	Light microscopy
LSP	Laser shock peening
MHP	Machine hammer peening
MOH	Maschinelles Oberflächenhämmern
MSLT	Mechanical surface layer treatment

MSV	Microstructural state variables
N_2O	Nitrous oxide
O-PS	Oxide polishing suspension for ductile metals
R3D4	Three-dimensional quadrilateral rigid element (ABAQUS)
RVE	Representative volume element
SEM	Scanning electron microscopy
SiC	Silicium carbid
SMAT	Surface mechanical attrition treatment
TRIP	Transformation induced plasticity
USD	United States Dollar
V2A	Versuchsschmelze 2 Austenit
VDI	VEREIN DEUTSCHER INGENIEURE (Association of German Engineers)
VHCF	Very high cycle fatigue
VUMAT	User subroutine in explicit FE simulation (ABAQUS)
WEF	World Economic Forum
W x H x L	Width x height x length
WKS	WHITTAKER-KOTEL'NIKOV-SHANNON

1 Introduction

Einleitung

Climate change is the world's greatest challenge. The *Global Risk Report 2020* of the World Economic Forum (WEF) lists for the first time exclusively five ecological global risks with the highest probability. These include extreme weather events, natural disasters and the loss of biodiversity [WORL20]. It is very likely that the current climate change is due to increased human activities since the middle of the 20th century [RAMA06]. The global warming associated with climate change is mainly explained by increased emissions of greenhouse gases such as carbon dioxide (CO_2), methane (CH_4) and nitrous oxide (N_2O) [KWEK18]. The contribution of CO_2 gases to global warming has risen continuously since 1979 and currently amounts to 66.1%, see Figure 1.1a [BUTL20].

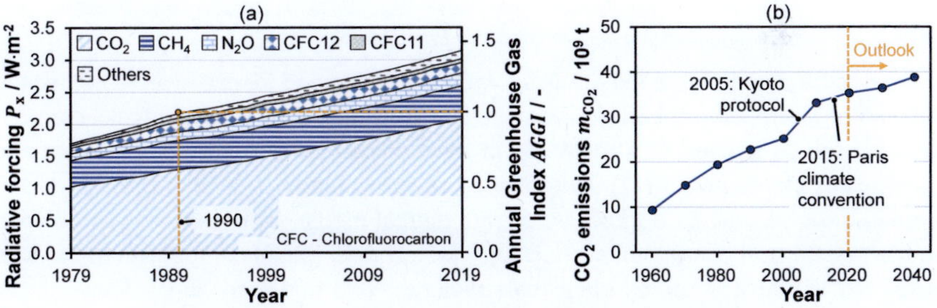

Figure 1.1: Development of the radiative forcing of long-lived greenhouse gases [BUTL20] (a) and evolution of worldwide CO_2 emissions [USDE19] (b)
Entwicklung der Strahlungswirkung von langlebigen Treibhausgasen (a) und Entwicklung der weltweiten CO_2-Emissionen (b)

In addition, it is predicted that global CO_2 emissions will continue to rise despite the climate agreements already concluded, see Figure 1.1b [USDE19]. The energy industry, especially the generation of electricity and heat, accounts for the largest share of 41% of total global emissions [IEA20]. This sector therefore has considerable potential to reduce CO_2 emissions and thus slow down climate change.

In Germany, CO_2 emissions from the energy sector have already been reduced by 33% compared to 1990. By 2030, the savings should be increased to at least 61% [DEUT20]. The expansion of renewable energies offers great potential for reducing CO_2 emissions. Currently, however, only a small proportion of the global energy demand is covered by renewable energies. Figure 1.2a shows that fossil energy sources such as coal, oil and natural gas are still predominantly used. However, the International Energy Agency (IEA) forecasts that the renewable energies will further expand depending on the political framework conditions in the coming years [IEA19]. In terms of share, a significant increase is expected especially in the area of wind, solar and bioenergy, see Figure 1.2b. Sustainable development scenarios predict that by 2040 renewable energies will cover almost 70% of the total global energy demand.

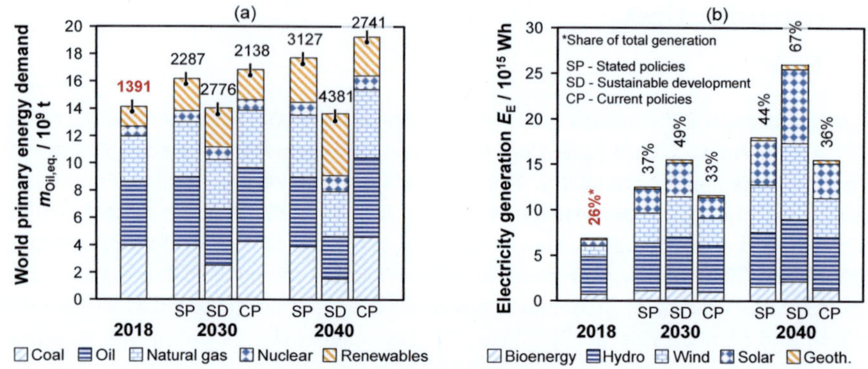

Figure 1.2: Current and projected development of global primary energy demand (a) and composition of renewable energies (b) [IEA19]
Aktuelle und prognostizierte Entwicklung des weltweiten Primärenergiebedarfs (a) und Zusammensetzung der erneuerbaren Energien (b)

In order to be able to meet the predicted increasing demand for renewable energies in the coming years, efficient energy conversion plants are required. The mostly complex systems are composed of a multitude of mechanical components, e.g. in wind and hydropower plants [WESS17]. Due to the prevailing loads, e.g. high pressures and temperatures as well as aggressive environmental media, high demands are placed on the mechanical components [SILB20]. Furthermore, global competition in the field of energy technology forces plant manufacturers and operators to continuously increase productivity and service life. The efficiency and sustainable functional safety of the plants is significantly influenced by the material and manufacturing properties of the individual components [WARE13].

Due to its material and processing properties, stainless steel is already being used in some areas of energy technology today. In addition to good corrosion resistance, this class of materials is characterized by a generally low yield strength, relatively high tensile strength and good ductility. Some stainless steel alloys also exhibit metastable behavior. This behavior allows an additional increase in strength due to phase transformations in the material during processing [PRAM15]. In hydroelectric power plants, stainless steel alloys are already used in precision gears, shafts or bolts. In biogas plants, they are used in dosing systems, cutting and stirring devices, heat exchangers, and filter and piping systems. Modern solar power plants use stainless steel alloys in central heat storage tanks with liquefied salt [WARE13]. However, a wider application of stainless steels in highly stressed components of energy technology is limited due to the insufficient mechanical strength and fatigue properties. A promising possibility to improve these properties is the use of forming processes. Since highly stressed technical components often fail at the surface, a targeted treatment of the surface layer already leads to significantly better component properties [PIPP06]. Improving the surface layer properties of highly stressed components from stainless steel by means of forming surface layer treatment therefore enables an expanded and load-adapted use of this steel grades in energy industry as well as an improvement in plant efficiency for renewable energies. Surface layer optimized stainless steel components thus contribute to CO_2 reduction.

Some studies have already demonstrated the potential of the well-known surface layer treatments shot peening and deep rolling to specifically improve the material properties relevant for increased fatigue strength, such as surface roughness [KONG19], work hardening [AMAN19], compressive residual stresses [NIKI07] and grain refinement [ALTE00]. A further, innovative surface layer treatment is machine hammer peening (MHP) [TRAU16A]. MHP is a cold micro-forging process in which the surface layer is machined at high frequency impacts by means of a mostly spherical plunger. Due to its process kinematics, MHP combines the advantages of shot peening and deep rolling. The MHP is divided into pneumatic [WIED11A], piezoelectric [LIEN17], mechanical [NETO20] and electrodynamic systems [LECH14] and is mostly adapted to machining centers or industrial robots. Despite some research work on the robot-guided electrodynamic MHP used in this work, there are still fundamental deficits in process kinetics, contact mechanisms and the surface and fatigue properties of hammered materials. MHP processing of austenitic stainless steels is also still largely unexplored.

Therefore, this thesis is concerned with the development of a comprehensive knowledge base of the process kinetics of electrodynamic MHP and the influence of industrial robots as MHP carrier system as well as the analysis of the predominant contact mechanisms and the resulting surface and fatigue properties of metastable AISI 304 stainless steel. By means of the metrological extension of an electrodynamic MHP system, cause-and-effect relationships between the forming energies and surface integrity and fatigue strength are identified for the first time and synthesized in the form of a design model for a future process layout. In this way, see Figure 1.3, this doctoral thesis makes a significant contribution to the knowledge of the robot-guided electrodynamic MHP of austenitic stainless steels, which in the future may lead to increased plant efficiency in the energy industry due to improved component properties.

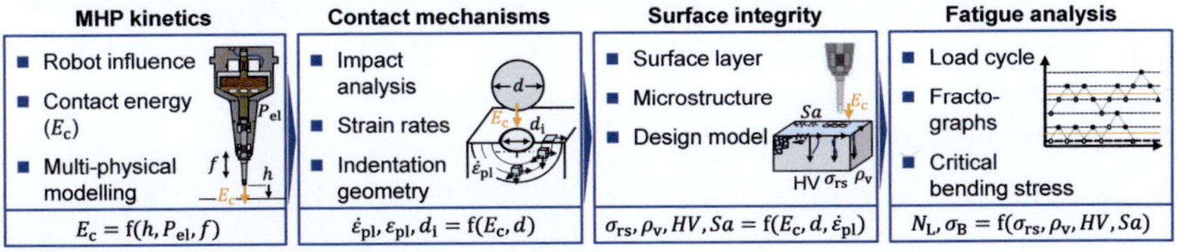

Figure 1.3: Structure of the doctoral thesis
Aufbau der Doktorarbeit

Einleitung

Der Klimawandel ist die größte Herausforderung für die Menschheit. Der *Global Risk Report 2020* des WORLD ECONOMIC FORUM (WEF) listet erstmals ausschließlich fünf ökologische globale Risiken mit der höchsten Wahrscheinlichkeit auf. Dazu zählen unter anderem Extremwetterereignisse, Naturkatastrophen und der Verlust der Biodiversität [WORL20]. Mit sehr hoher Wahrscheinlichkeit ist ebenjener aktueller Klimawandel auf verstärkte menschliche Aktivitäten seit Mitte des 20. Jahrhunderts zurückzuführen [RAMA06]. Die mit dem Klimawandel einhergehende Erderwärmung wird vor allem durch den erhöhten Ausstoß an Treibhausgasen wie Kohlenstoffdioxid (CO_2), Methan (CH_4) und Distickstoffmonoxid (N_2O) erklärt [KWEK18]. Der Beitrag der CO_2-Gase zur Erderwärmung ist dabei seit 1979 kontinuierlich gestiegen und beträgt aktuell 66,1 %, siehe Abbildung 1.1a [BUTL20].

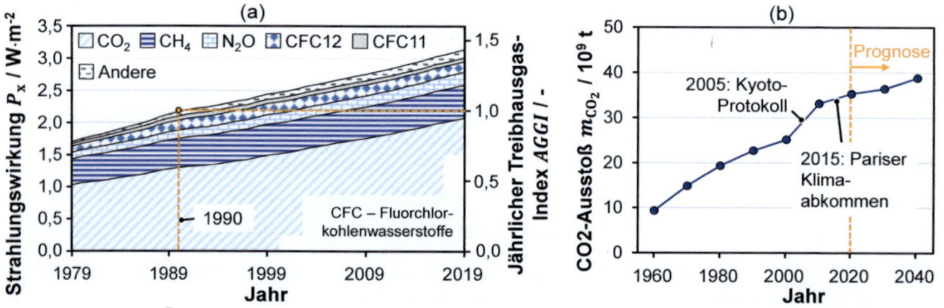

Abbildung 1.1: Entwicklung der Strahlungswirkung von langlebigen Treibhausgasen [BUTL20] (a) und Entwicklung der weltweiten CO2-Emissionen [USDE19] (b)
Development of the radiative forcing of long-lived greenhouse gases (a) and evolution of worldwide CO_2 emissions (b)

Darüber hinaus wird prognostiziert, dass der weltweite CO_2-Ausstoß trotz bereits geschlossener Klimaabkommen weiter steigen wird, siehe Abbildung 1.1b [USDE19]. Den größten Anteil an den weltweiten Gesamtemissionen verursacht dabei mit 41 % die Energiewirtschaft, besonders die Elektrizitäts- und Wärmeerzeugung [IEA20]. In diesem Sektor besteht daher ein erhebliches Potenzial, den CO_2-Austoß zu senken und somit den Klimawandel zu verlangsam.

In Deutschland wurden bis zum heutigen Tag die CO_2-Emissionen aus dem Bereich Energiewirtschaft bereits um ca. 33 % im Vergleich zu 1990 gesenkt. Bis 2030 sollen die Einsparungen auf mindestens 61 % erhöht werden [DEUT20]. Großes Potenzial, die CO_2-Emissionen zu verringern, bietet dabei der Ausbau erneuerbarer Energien. Aktuell wird jedoch nur ein geringer Anteil des weltweiten Energiebedarfs durch erneuerbare Energien gedeckt. Abbildung 1.2a zeigt, dass zurzeit überwiegend noch fossile Energieträger wie Kohle, Öl und Erdgas zum Einsatz kommen. Die INTERNATIONALE ENERGIEAGENTUR (IEA) prognostiziert allerdings in den kommenden Jahren einen von den politischen Rahmenbedingungen abhängigen Ausbau der erneuerbaren Energien [IEA19]. Anteilig wird besonders im Bereich der Wind-, Sonnen- und Bioenergie ein deutlicher Anstieg erwartet, siehe Abbildung 1.2b. Nachhaltige Entwicklungsszenarien rechnen damit, dass zum Jahre 2040 die erneuerbaren Energien knapp 70 % des weltweiten Gesamtenergiebedarfs decken.

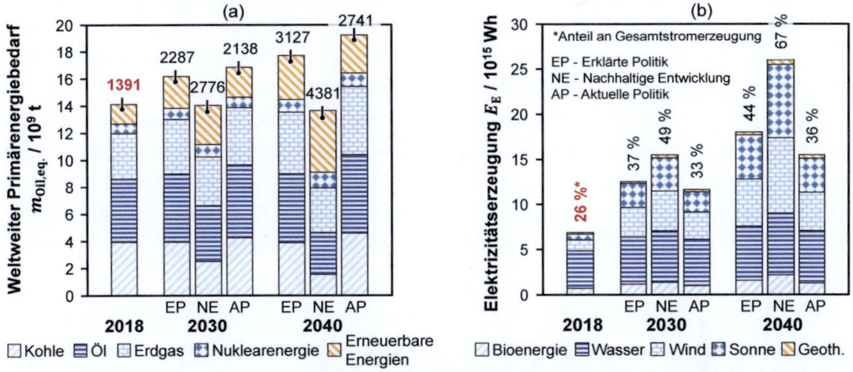

Abbildung 1.2: Aktuelle und prognostizierte Entwicklung des weltweiten Primärenergiebedarfs (a) und Zusammensetzung der erneuerbaren Energien (b) [IEA19]
Current and projected development of global primary energy demand (a) and composition of renewable energies (b)

Um die prognostizierte, steigende Nachfrage nach erneuerbaren Energien in den kommenden Jahren bedienen zu können, werden leistungsfähige Anlagen zur Energieumwandlung benötigt. Die zumeist komplexen Anlagen setzen sich aus einer Vielzahl von mechanischen Bauteilen zusammen, bspw. in Wind- und Wasserkraftanlagen [WESS17]. An die mechanischen Bauteile werden aufgrund der vorherrschenden Bedingungen, bestehend aus komplexen Belastungskollektiven und hohen Temperaturen sowie aggressiven Umgebungsmedien, hohe Anforderungen gestellt [SILB20]. Weiterhin zwingt der globale Wettbewerb im Bereich der Energietechnik die Anlagenhersteller und -betreiber zu einer fortwährenden Produktivitäts- und Standzeiterhöhung, was wiederrum die Anforderungen an die Bauteile erhöht. Die Effizienz sowie nachhaltige Funktionssicherheit der Anlagen wird dabei maßgeblich durch die Werkstoff- und Bearbeitungseigenschaften der einzelnen Bauteile beeinflusst [WARE13].

Aufgrund seiner Werkstoff- und Bearbeitungseigenschaften wird rostfreier Edelstahl heutzutage bereits in einigen Bereichen der Energietechnik eingesetzt. Neben einer guten Korrosionsbeständigkeit zeichnet sich diese Werkstoffklasse durch eine zumeist niedrige Streckgrenze, eine relativ hohe Zugfestigkeit sowie gute Zähigkeit aus. Einige Edelstahllegierung weisen zusätzlich ein metastabiles Verhalten auf. Dieses Verhalten ermöglicht eine weitere Festigkeitssteigerung aufgrund von Phasenumwandlungen im Werkstoff während der Bearbeitung [PRAM15]. In Wasserkraftwerken kommen Edelstahllegierungen bereits heute in Präzisionszahnrädern, Wellen oder Bolzen zum Einsatz. In Biogasanlagen werden sie in Dosieranlagen, Schneid- und Rührwerken, Wärmetauschern sowie Filter- und Rohrleitungen eingesetzt. Moderne Solarkraftwerke nutzen Edelstahllegierungen in zentralen Wärmespeichertanks mit verflüssigtem Salz

[WARE13]. Eine breitere Anwendung rostfreier Edelstähle in hochbelasteten Bauteilen der Energietechnik wird jedoch aufgrund der zum Teil unzureichenden mechanischen Festigkeits- und Ermüdungseigenschaften begrenzt. Eine vielversprechende Möglichkeit diese Eigenschaften gezielt zu verbessern, besteht in dem Einsatz von umformenden Fertigungsverfahren. Da hochbelastete technische Bauteile häufig an der Oberfläche versagen, führt eine gezielte umformtechnische Randzonenbearbeitung bereits zu signifikant besseren Bauteileigenschaften [PIPP06]. Die Verbesserung der Randzoneneigenschaften von hochbeanspruchten Bauteilen aus nichtrostendem Edelstahl durch eine formgebende Randzonenbearbeitung ermöglicht daher einen erweiterten und belastungsangepassten Einsatz dieser Stahlsorten in der Energiewirtschaft sowie eine Verbesserung der Anlageneffizienz für erneuerbare Energien. Randzonenbearbeitete Edelstahlbauteile tragen somit zur CO_2-Reduzierung bei.

Einige Arbeiten haben bereits das Potenzial der bekannten mechanischen Randzonenbearbeitungsverfahren (RZBV) Kugelstrahlen und Festwalzen, welche die für eine erhöhte Ermüdungsfestigkeit relevanten Werkstoffeigenschaften wie Oberflächenrauheit [KONG19], Kaltverfestigung [AMAN19], Druckeigenspannungen [NIKI07] und Kornfeinung [ALTE00] gezielt verbessern, nachgewiesen. Ein weiteres, innovatives RZBV ist das maschinelle Oberflächenhämmern (MOH) [TRAU16B]. Das MOH ist ein Kaltmikroschmiedeverfahren, welches metallische Randzonen hochfrequent mittels eines zumeist sphärischen Stößels bearbeitet. Dabei vereint das MOH aufgrund seiner Prozesskinematik die Vorteile des Kugelstrahlens und des Festwalzens in sich. Unterschieden wird das MOH hinsichtlich des zugrundeliegenden Wirkprinzips in pneumatische [WIED11A], piezoelektrische [LIEN17], mechanische [NETO20] und elektrodynamische Systeme [LECH14]. Zumeist werden die MOH-Werkzeuge über ein Bearbeitungszentrum oder einen Industrieroboter geführt. Trotz einiger Forschungsarbeiten zum in dieser Arbeit eingesetzten robotergeführten elektrodynamischen MOH existieren weiterhin grundlegende Wissensdefizite im Bereich der Prozesskinetik, der Kontaktmechanismen sowie der resultierenden Oberflächen- und Ermüdungseigenschaften gehämmerter Werkstoffe. Die MOH-Bearbeitung von rostfreien Edelstählen ist ebenfalls noch größtenteils unerforscht.

Gegenstand dieser Arbeit ist daher neben dem Aufbau einer umfangreichen Wissensbasis der Prozesskinetik des elektrodynamischen MOH sowie dem Einfluss von Industrierobotern als MOH-Trägersystem auch die Analyse der vorherrschenden Kontaktmechanismen sowie die resultierenden Oberflächen- und Ermüdungseigenschaften von metastabilem austenitischen AISI 304-Edelstahl, siehe Abbildung 1.3. Durch die messtechnische Erweiterung eines elektrodynamischen MOH-Systems werden erstmalig Ursache-Wirkungzusammenhänge zwischen den eingebrachten Energien und der resultierenden Oberflächenintegrität sowie der Ermüdungsfestigkeit identifiziert und in Form eines Gestaltungsmodells für eine künftige Prozessauslegung synthetisiert. Auf diese Weise liefert diese Dissertation einen wesentlichen Wissensbeitrag zum robotergeführten elektrodynamischen MOH von rostfreien austenitischen Edelstählen, was zukünftig in einer verbesserten Bauteileigenschaft und somit gesteigerten Anlageneffizienz im Bereich der Energieindustrie führen kann.

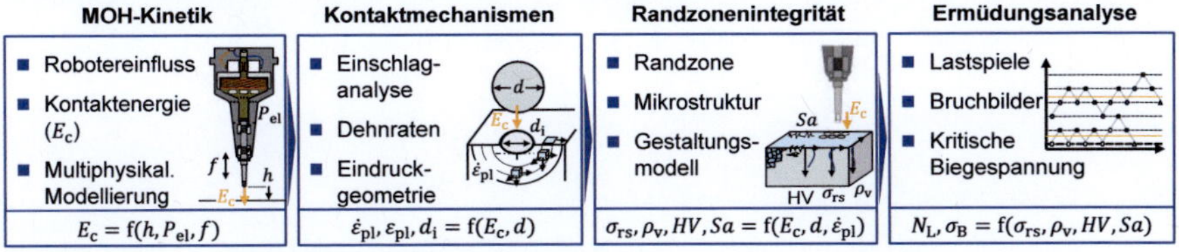

Abbildung 1.3: Aufbau der Doktorarbeit
Structure of the doctoral thesis

2 State of the art in research and industry

Stand der Technik in Forschung und Industrie

The subject of this thesis is the analysis of the influence of electrodynamic machine hammer peening (MHP) on the surface layer mechanisms of austenitic stainless steel and the workpiece fatigue strength resulting from the adapted surface integrity. The chosen approach is to generate a comprehensive understanding of the multiphysical system behavior of electrodynamic MHP. The systemic knowledge allows to explain the forming conditions occurring in workpiece contact and to derive cause-and-effect relationships with the resulting surface integrity. Thereby, the multiple material-physical mechanisms during the forming of metastable austenitic stainless steels require special consideration. By knowing the elasto-plastic mechanisms in the workpiece surface layer because of defined forming work introduced, the remaining surface integrity can be specifically adjusted and its influence on the fatigue properties described.

The basics for the understanding of the above-mentioned contents dealt within this research study are worked out in Chapter 2. Chapter 2.1 first discusses, among other things, the basic principles of materials science as well as the transformation mechanisms and deformation behavior of austenitic stainless steels. In Chapter 2.2, fundamental aspects and mechanisms of action of mechanical surface layer treatment (MSLT) processes are explained. Based on this, the process of MHP analyzed in this thesis and the mechanical contact mechanisms resulting from the forming energy involved are presented in Chapter 2.3. Finally, the state of the art is summarized in Chapter 2.4 and an economic and scientific problem statement is derived.

2.1 Austenitic stainless steels

Austenitische rostfreie Stähle

Stainless steels are a group of steel grades with high corrosion and acid resistance, but poor thermal and electrical conductivity compared to copper. As early as 1821, BERTHIER first researched their production [LOWE17]. In Germany, the first patents were applied by MAURER and STRAUSS of the KRUPP RESEARCH INSTITUTE in 1912 [RASC12]. According to today's European standard EN 10088-1:2014, stainless steels are steels characterized by a chrome (Cr) content of more than 10.5% and a carbon (C) content not exceeding 1.2% dissolved in austenitic or ferritic solid solution [DIN14A]. Stainless steel is used in countless applications in the automotive and aerospace industry, the construction and food industries, the chemical industry and medical technology [DEW16]. Furthermore, stainless steel will make an important contribution to energy production in the next years [ISSF16]. The global production volume of stainless steel has risen continuously in the past, with China accounting for by far the largest share with more than 50%, see Figure 2.1a [ISSF20]. An production increase in China is expected for stainless steel by 2027, see Figure 2.1b [GVR20].

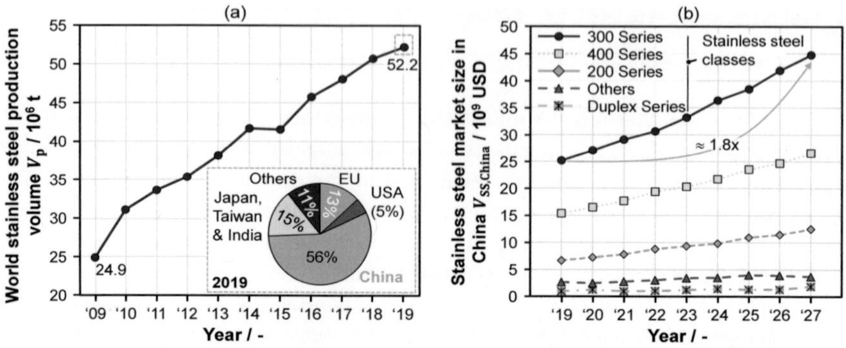

Figure 2.1: Development of global stainless steel production from 2009 to 2019 [ISSF20] (a) and forecast of stainless steel market development in China [GVR20] (b)
Entwicklung der weltweiten Edelstahlproduktion von 2009 bis 2019 (a) und Prognose der Entwicklung des Edelstahlmarktes in China (b)

Austenitic stainless steels represent the most important class of stainless steels. They account for three quarters of the global production. They have a low yield strength, high ductility, very good corrosion properties and pronounced work-hardening behavior. The metastable austenitic stainless steel X5CrNi18-10 (DIN 1.4301, AISI 304) is the most widely used stainless steel with 33% of worldwide production volume and belongs to the 300 series [MÜLL11]. As shown in Figure 2.1b, a strong increase in 300 series stainless steels in China is predicted [GVR20]. The pronounced work-hardening behavior of metastable austenitic stainless steels is the key parameter, which enables the targeted adjustment of the component properties over a wide range. Therefore, the underlying mechanisms responsible for this behavior are explained in more detail in the following subsections. Based on this, the resulting challenges and current knowledge deficits for the targeted property adjustment of metastable austenitic stainless steels are formulated.

2.1.1 Austenite stability

Austenitstabilität

The basic principle for the production of austenitic stainless steels is the stabilization of the iron phase γ-austenite [HEIN04]. Since the γ-austenite is unstable at room temperature, it must be chemically or mechanically stabilized. SCHEIL describes this in detail [SCHE16]. The γ-austenite stability is influenced by the alloy composition, the temperature T, the stress state, the γ-austenite grain size d_γ, the stacking fault energy γ_{SF} and the strain rate $\dot{\varepsilon}_{tot}$. Especially the alloy composition has a fundamental importance for the γ-austenite stability [HEIN04].

The alloying elements chrome (Cr) and nickel (Ni) have the greatest influence on the microstructure of stainless steels. By adding Cr, a protective, passivating chrome oxide layer is formed in a subsequent oxygen atmosphere [RIED10]. By adding further alloying elements such as nickel (Ni), carbon (C), manganese (Mn), molybdenum (Mo), titanium (Ti) and niobium (Nb), the mechanical properties of the material are specifically adjusted and the corrosion resistance further improved [ISER08]. Depending on the influence of the individual alloying elements on the favorable formation of α- or δ-ferrite (Cr, Mo, Nb) or γ-austenite (Ni, C, Mn) phases, these are referred to as ferrite or austenite formers. An overview can be found in [ROTH12]. Using the so-called chrome (Cr_{eq}) and nickel equivalents (Ni_{eq}), the austenitic or ferritic stabilizing effect of the alloying elements can be estimated by means of an empirically determined effective sum calculation. In his work, SCHOSS presents the Equations (2.1) and (2.2) for determining the equivalent values [SCHO01]:

$$Cr_{\mathrm{eq}} = \%Cr + \%Mo + 1.5\%Si + 0.5\%Nb + 2\%Ti \quad \textbf{(2.1)}$$

$$Ni_{\mathrm{eq}} = \%Ni + 30\%C + 0.5\%Mn \quad \textbf{(2.2)}$$

The SCHAEFFLER diagram in Figure 2.2a shows the stable phases in stainless steels which form as a function of Cr_{eq} and Ni_{eq} [SCHA49]. In addition to austenitic phases (A), stainless steels with ferritic (F), martensitic (M) and duplex microstructures (A+F) can also be produced by specific alloy adjustment [KRAU10]. The austenitic stainless steel AISI 304 examined in this work is, after calculation of Cr_{eq} and Ni_{eq}, at the limit (metastable) of a purely austenitic microstructure with potentially small amounts of ferrite and martensite (orange point).

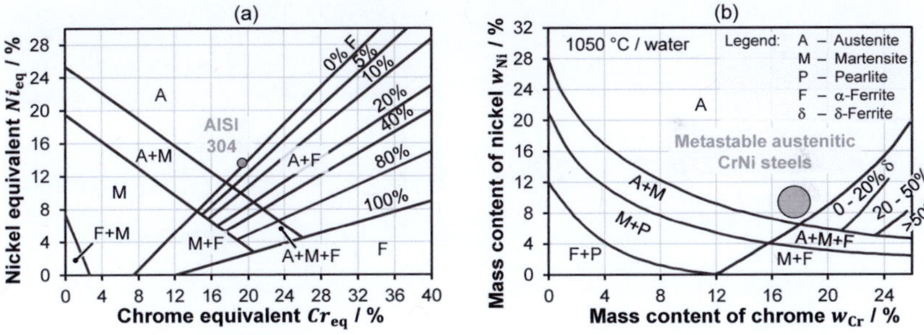

Figure 2.2: SCHAEFFLER diagram for phase determination in stainless steels [SCHA49] (a) and MAURER diagram for phase determination in CrNi steels [HONE81] (b)
SCHAEFFLER-Diagramm zur Phasenbestimmung in rostfreien Edelstählen (a) und MAURER-Diagramm zur Phasenbestimmung in CrNi-Stählen (b)

The influence of Cr and Ni contents in stainless CrNi steels can be shown using the microstructure diagram according to MAURER, see Figure 2.2b [HONE81]. An austenitic microstructure of AISI 304 is set after a standard solution heat treatment to 1,050 °C under inert gas atmosphere and subsequent quenching in water [WEBE12]. Due to the high costs of the alloying element Ni, the aim is to achieve an austenitic microstructure for most stainless CrNi steels with Cr contents of 18% and Ni contents of 8%. Cr contents above 18% lead to an increased formation of δ-ferrite [SCHU09]. After quenching, there is a state of equilibrium, i.e. a metastable γ-austenite microstructure, in which diffusion-controlled as well as diffusion-free microstructural changes can occur. The underlying martensitic transformation is explained in the following subsection.

2.1.2 Martensitic transformation

Martensitumwandlung

A martensitic transformation describes a diffusion-free, lattice-distorting change in the metallic crystal lattice structure. This structural change is dominated by lattice-variant and lattice-invariant shear [HERP00]. According to OLSON and COHEN, a martensitic transformation from the metastable γ-austenite can be induced thermally as well as stress and strain (deformation) induced, see Figure 2.3a [OLSO72].

Thermodynamic approach and chemical composition

By means of the three above-mentioned approaches, martensitic nuclei can be formed and kinetic obstacles can be overcome to realize growth of these martensitic nuclei. Thereby, the free enthalpy difference ΔG between the γ-austenite and martensite phases acts as the driving force of the martensitic transformation [LLEW97]. Stress-induced martensite occurs at those nuclei where thermally induced martensite formation is initiated, supplemented by a thermodynamic stress effect. Thermally induced martensite is formed by high cooling rates of the material to the so-called martensite starting temperature M_s, far below room temperature, see Figure 2.3a [HORN15].

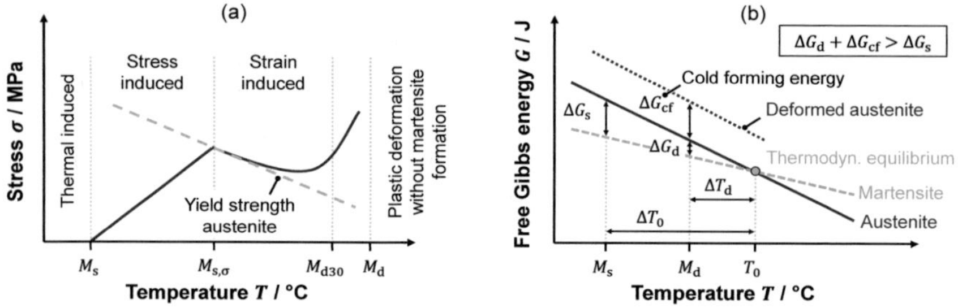

Figure 2.3: Critical stress for initiation of martensitic transformation as function of temperature [OLSO72] (a) and free GIBBS energy of austenite and martensite as function of temperature [PATE53] (b)

Kritische Spannung zur Einleitung der Martensitumwandlung in Abhängigkeit von der Temperatur (a) und freie GIBBS-Energie von Austenit und Martensit in Abhängigkeit von der Temperatur (b)

Between the M_s and M_d temperatures, stress-induced martensite is formed under elastic deformation [BOLL70]. Strain-induced martensite is formed on nuclei from the plastic deformation and can be caused by sufficiently high stresses at room temperature (here M_d) [MÜLL11]. The M_s temperature is determined by the chemical composition and the γ-austenite grain size. M_s is often calculated according to the EICHEL-HULLMANN-HULL relationship, see Equation (2.3) [EICH53]. Further equations are listed by HEINEMANN [HEIN04].

$$M_s = 1350 - 1665 \cdot (\%C + \%N) - 28 \cdot \%Si - 33 \cdot \%Mn - 42 \cdot \%Cr - 61 \cdot \%Ni \quad \textbf{(2.3)}$$

In addition to the M_d temperature, the M_{d30} temperature is often used because it is easier to determine. At the M_{d30} temperature, 50 vol.-% martensite is formed after a deformation of 30% [NOHA77]. Thereby, larger γ-austenite grains reinforce the effect of the α'-martensite transformation [VARM94]. Figure 2.3b shows that a transformation from an austenitic to a martensitic microstructure does not occur only by cooling the material with high cooling rates below T_0 (thermodynamic equilibrium). Undercooling with ΔT is required to allow spontaneous transformation. At room temperature, the necessary enthalpy difference ΔG_s is provided for example by sufficient cold forming work (ΔG_{cf}). In addition to a thermodynamic view of the martensitic transformation, the occurring transformation mechanisms can also be described crystallographically.

Phenomenological and crystallographic description

When the martensite start temperature M_s is reached, martensite formation begins suddenly. The diffusion-free transformation, which takes place within milliseconds, results in a successive folding over of the old into the new crystal structure. The folding over leads to the formation of needle-, plate- and lancet-shaped areas [BLEC01]. Martensite needles essentially consist of twins [WEIS94]. In alloyed steels, ε-martensite with hexagonal close-packed (hcp) and α'-martensite with body-centered cubic (bcc) crystal structure occurs. The increase in volume caused by α'-martensite transformation is an alloy-dependent variable (mainly C) [RUGE07] and is stated in the literature as approximately 2% [SURE04]. Due to its higher strength compared to the γ-austenite and ε-martensite phase, the α'-martensite phase is of great importance for the mechanical properties of austenitic stainless steels. Since ε- and α'-martensite form via different transformation paths, these are explained in the following.

In a former γ-austenite grain several thin ε-martensite plates usually accumulate parallel to packages with grain boundaries and twins limiting their growth. The formation of α'-martensite usually occurs at intersections of the ε-martensite plates. For this purpose, SCHUMANN presented three different transformation paths for the formation of ε- or α'-martensite: γ-α' (1), γ-ε-α' (2) and γ-ε (3) [SCHU75]. α'-martensite, which is formed directly from γ-austenite, appears block-shaped or longitudinal [HENN76] and causes greater hardening [SPRI06]. The α'-martensite formed via (2) appears lath-shaped or needle-shaped [KELL65]. To explain the direct transformation path from γ-austenite to α'-martensite, the BAIN model is usually used [BAIN24]. However, the crystal orientation relationships resulting from this direct transformation according to NISHIYAMA and WASSERMANN [NISH78] and KURDJUMOV and SACHS [KURD30] have been proven more frequently. Recently, WANG ET AL. presented further formation mechanisms for stress- and strain-induced martensite formation [WANG20B].

ε-martensite is difficult to distinguish from stacking faults, as both crystal defects are caused by shear. OLSEN and COHEN therefore summarized these two strain-induced defects together with twins and dislocations under the concept of *shear bands* [OLSO75]. It has already been shown many times that in austenitic stainless steels,

α'-martensite is formed preferentially at crossing points of shear bands [SCHO01]. Detailed explanations of the martensite formation mechanisms and nucleation points can be found in [MÜLL11].

Metal-physical explanation

The partly contradictory statements on possible α'-martensite formation mechanisms in the literature have in common that the stacking fault energy γ_{SF} is of great importance for the mechanisms of martensite formation [MÜLL11]. A low stacking fault energy γ_{SF} favors the formation of stacking faults, deformation twins or ε-martensite. The stacking fault energy of austenitic stainless steels is usually in the range of γ_{SF} = 20 mJ/m² [LU16]. The following empirical Equation (2.4) for calculating γ_{SF} was proposed by SCHRAMM and REED [SCHR75], further approaches were formulated by RHODES and THOMPSON [RHOD77] and DAI ET AL. [DAI02].

$$\gamma_{\mathrm{SF}} = -53 + 6.2 \cdot \%Ni + 0.7 \cdot \%Cr + 3.2 \cdot \%Mn + 9.3 \cdot \%Mo \qquad \textbf{(2.4)}$$

VÖHRINGER and MACHERAUCH explained the formation of preformed martensite nuclei by dislocation reactions of already existing dislocations [VÖHR77]. The displacements of dislocations required for transformation take place along the most densely packed planes, similar to plastic deformation by sliding. The stacking faults resulting from dislocation splitting are interpreted as the starting point for nucleation [GOTT07]. Further dislocation movement increases the nuclei size. The martensite formation starts in the most favorable position with the lowest required stresses, often the {111}-closest packed slip plane in fcc (face centered cubic) crystals. This leads to an increase in the structural defect density [DASH63]. SUZUKI ET AL. [SUZU76] as well as FANG and DAHL [FANG91] also formulated the internal stresses resulting from dislocation processes as a prerequisite for the mechanical driving force for martensite transformation. BARKLEY summarized the structural change during deformation of metastable austenitic CrNi steels in six stages from dislocation movement in γ-austenite to the formation of deformation twins in martensite [BARC65].

In the previous sections, the martensite formation mechanisms have been described in detail. Since this work investigates a cold forming problem, strain-induced martensite transformation is of particular interest. Therefore, the influencing variables on strain-induced martensite formation are explained below.

Influencing factors on strain-induced martensite formation

The strain-induced martensite formation is significantly determined by the chemical composition, temperature T, strain $\varepsilon_{\mathrm{tot}}$, strain rate $\dot{\varepsilon}_{\mathrm{tot}}$ and grain size of the γ-austenite d_{γ} [TALO07]. In the following, the influencing variables plastic strain $\varepsilon_{\mathrm{pl}}$ and plastic strain rate $\dot{\varepsilon}_{\mathrm{pl}}$, which are the most relevant for this work, are dealt with in particular.

Strain state

Some studies report that under tensile stress, strain-induced martensite formation occurs more strongly than under compressive stress [POWE58]. IWAMOTO ET AL. found out for AISI 304 steel that the transformation rate at low strains is higher at compressive stress than at tensile stress [IWAM98]. SHRINIVAS ET AL. explained that more α'-martensite was generated in AISI 316 steel in the cold-rolled state than in the uniaxial tensile test and attributed this to the large number of activated sliding systems [SHRI95]. SPENCER ET AL. observed that α'-martensite is formed directly from the γ-phase at sufficiently high strains [SPEN09]. MURR ET AL. [MURR82] concluded that multiaxial strain states lead to increased martensite formation, see Figure 2.4a.

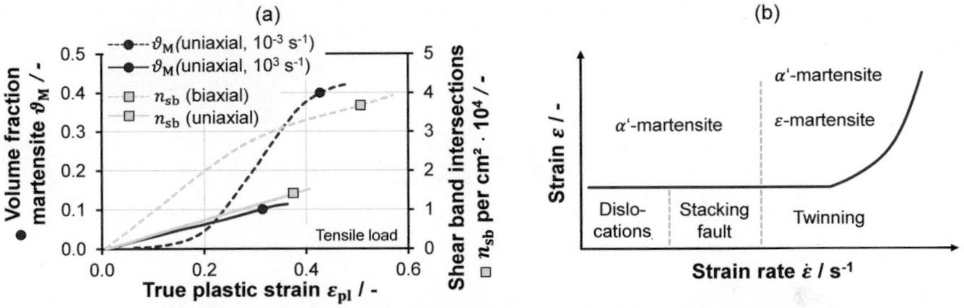

Figure 2.4: Influence of different strains [MURR82] and strain rates [HECK82] on the martensite volume fraction and number of shear band intersections (a) and microstructural evolution as function of strain and strain rate according to FERREIRA ET AL. [FERR04] (b)
Einfluss von verschiedenen Dehnungszuständen sowie Dehnraten auf den Martensitvolumenanteil und die Anzahl Scherbandkreuzungen (a) und Gefügeentwicklung in Abhängigkeit von der Dehnung und Dehnrate nach FERREIRA ET AL. (b)

Strain rate

The influence of the strain rate must be divided into a direct and indirect component. In the case of the indirect component, very high strain rates can lead to adiabatic material heating, a lower chemical driving force and thus inhibited martensite formation [SANT06]. HECKER ET AL. explained this by the increased formation of irregular shear band arrangements at higher strain rates and thus a reduced probability of the formation of α'-martensite nuclei of a critical size [HECK82]. Other studies reported increased shear band formation at higher strain rates, see Figure 2.4a. This results in more α'-martensitic nuclei and therefore more α'-martensite at low strain $\varepsilon_{pl} < 0.25$ during tensile loading [HECK82]. LEE and LIN investigated the effects of strain rates $\dot{\varepsilon}_{pl} = 8 \cdot 10^2 \ldots 4.8 \cdot 10^3$ s^{-1} on α'-martensite formation in upsetting tests with strains $\varepsilon_{pl} = 0.05 \ldots 0.3$. They found that the α'-martensite content increased with increasing strain rate [LEE00]. DE ET AL. [DE06] and TALONEN [TALO07] showed that at low strains, when the self-heating of the material is still low, increased martensite formation is observed in combination with high strain rates. HUANG ET AL. explained the influence of dislocation densities and their interactions as the reason for the influence of the strength properties in addition to the strain-induced martensite transformation [HUAN89]. FERREIRA ET AL. elaborated a schematic diagram (see Figure 2.4b) that shows the various characteristics of AISI 304 microstructure as a function of strain and strain rate. They pointed out that an increasing strain rate favors the formation of deformation twins. In combination with sufficiently high strains, α'-martensite is formed.

The above paragraphs have shown that different mechanisms of martensite transformation are activated during the transformation of metastable austenite depending on the strain states and strain rates. However, the influence of the technically relevant α'-martensite phase on the material properties must be determined. Therefore, the effects of the α'-martensite phase on the strength properties of austenitic stainless steels are discussed in the following.

Effect of α'-martensite on material properties of austenitic stainless steel

With increasing content of deformation martensite, the yield stress of austenitic stainless steels increases due to the higher strength of the martensite [ANGE54]. Furthermore, the strain-induced α'-martensite transformation is accompanied with an increase in work-hardening rate and an increase of tensile strength, see Figure 2.5 [BYUN04].

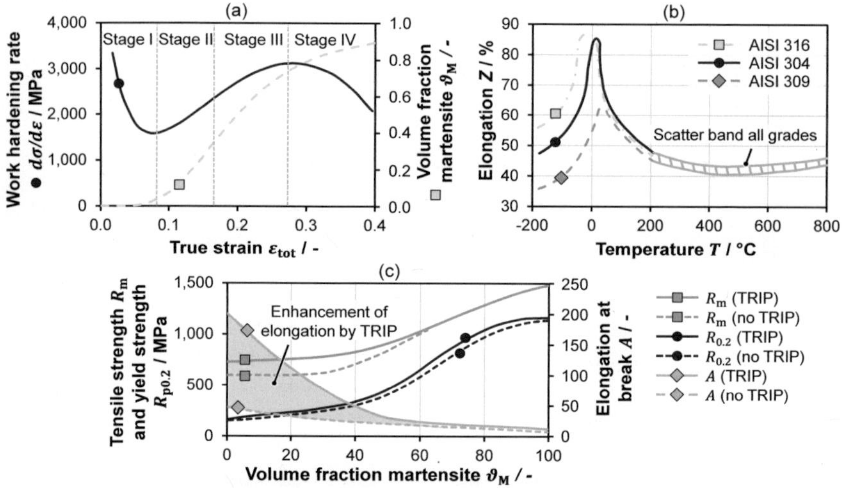

Figure 2.5: Development of work hardening rate for low strain rates and room temperature [TALO07] (a), elongation behavior of selected austenitic steels [SAND69] (b) and influence of TRIP effect on mechanical properties [TSUC00] (c)

Verlauf der Kaltverfestigungsrate für niedrige Dehnraten und bei Raumtemperatur (a), Dehnungsverhalten von ausgewählten austenitischen Stählen (b) und Einfluss des TRIP-Effektes auf die mechanischen Eigenschaften (c)

The curve of work hardening as a function of strain shows a characteristic minimum before reaching a maximum, which has been related to the formation of the α'-martensite phase, and is divided into four different phases. A detailed description of the four phases can be found in [TALO07]. ZACKAY ET AL. summarized the various effects of α'-martensite transformation as TRIP effect (TRansformation-Induced Plasticity), in which both an increase in strength and ductility can be determined [ZACK67]. The ductility is largely determined by the martensite formation rate and the point of maximum work-hardening rate (α'-martensite transformation), see Figure 2.5b [LIVI77]. The positive influence of α'-martensite formation on strength parameters such as yield strength $R_{p0.2}$ and tensile strength R_m as well as the ductility (e.g. elongation at break A) of an austenitic stainless steel is shown in Figure 2.5c [TSUC00].

Local martensite formation in areas of high plastic deformation prevents or hinders necking and thus delays fracture [HUAN89]. Nonetheless, it is necessary to further distinguish between the strengthening mechanism by means of α'-martensit transformation and strengthening effects due to creation of dislocation substructures in the γ-austenite phase [BYUN04]. MÜLLER-BOLLENHAGEN showed that there is a linear relationship between the root of the dislocation density and the yield stress up to a α'-martensite content greater than 30%. Up to this limit, at least the strain hardening during the strain-induced martensite transformation is mainly determined by the dislocation density of the austenite phase in AISI 304 stainless steel [MÜLL11]. The mechanisms of plastic deformation, which is dominated by dislocation evolution, and the resulting microstructural changes are explained below.

2.1.3 Deformation behavior of austenitic stainless steels

Verformungsverhalten von austenitischen rostfreien Stählen

The generation and movement of dislocations, i.e. two-dimensional lattice defects, significantly influence plastic deformation. Idealized dislocation types are screw and edge dislocations. Dislocations move along different slip systems during plastic deformation. The dislocation density describes the length of the dislocation lines per unit volume. Plastic deformation results in an increased dislocation density. A detailed explanation can be found in [GOTT14]. As the dislocation development during cold forming is both material and load dependent, the mechanisms relevant for austenitic stainless steels are examined in more detail.

Mechanism of plastic deformation and metal physical modelling

In austenitic stainless steels, dislocations tend to split into partial dislocations during forming. This reduces the energy required to move the dislocation. A stacking fault with a certain splitting width is formed between the partial dislocations, which determines the stacking fault energy [MÜLL11]. A low stacking fault energy γ_{SF} as in austenitic stainless steels is characterized by a large splitting width. This often results in dislocation movements without transverse slip.

During cold forming, the shape of initial grains changes and often more grain boundaries are created by the incorporation of some of the continuously created dislocations. The increase in grain boundary area is significantly influenced by the strain state ε_{pl} and strain rate $\dot{\varepsilon}_{pl}$ [CRUZ13]. A second microstructural change during plastic deformation is the creation of an internal structure within the initial grains by means of dislocation creation, movement and accumulation [CLAY06]. HUMPHREYS and HATHERLY described the hierarchy of microstructure in polycrystalline metal forming by slip.

Thereby, subgrains or cells (low angle grain boundaries) consist of dislocations in different configurations and subdivide the initial grains. On larger scales, dislocation arrangements with different orientations inside one grain form deformation bands. The continuing of the deformation bands leads to shear bands across a few grains, see Figure 2.6 [HUMP04].

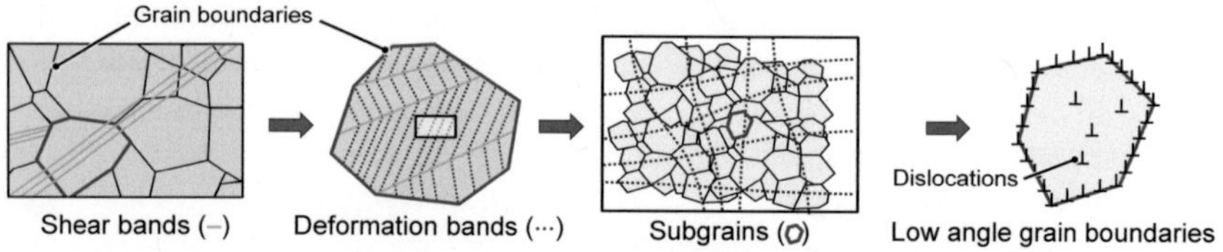

Figure 2.6: Microstructure hierarchy in polycrystalline metal forming by slip mechanism according to HUMPHREYS and HATHERLY [HUMP04]
Hierarchie der Mikrostruktur während der Umformung polykristalliner Metalle nach HUMPHREYS und HATHERLY

The deformation behavior of metastable austenite phases can be described most easily analytically by means of modified, empirical descriptions of the classical stress-strain relationships. These *empirical constitutive models* often describe the yield stress as an explicit function of accumulated plastic strain, strain rate and temperature [MOTA19]. Some common models are the ones of FOLLANSBEE and KOCKS [FOLL88], JOHNSON and COOK [JOHN83] and SUNG ET AL. [SUNG10]. In most cases of forming of metastable austenitic stainless steel, an additive contribution of austenite and martensite is assumed, neglecting the contribution of ε-martensite. HÄNSEL using the HOCKETT-SHERBY hardening law [HÄNS98] as well as TOMITA and IWAMOTO [TOMI95] described such classical approaches in their work. A disadvantage of these empirical models is the high number of model constants, which are without any physical meaning and have to be adapted to experimental data [MOTA19].

Further approaches describe the mechanical behavior of metastable austenitic stainless steels by means of microstructural changes during forming (e.g. changes in dislocation density and grain size). The basis of these *microstructural constitutive models* is the classical relationship between dislocation density and yield stress [MÜLL11]. The knowledge about the microstructural processes including dislocation mechanisms allows for the formulation of thermo-micro-mechanical constitutive equations [CHAN18]. Furthermore, these models can be used to model the mechanical responses (plastic deformation) as a result of the main microstructural phenomena in a wide range of strain rates and deformation modes [MOTA19]. Thereby, the approach of *discrete dislocation dynamics (DDD)* in representative volume elements (RVE) allows for a detailed study of dislocation processes, such as arrangements, configurations and interactions [PAPA17]. Nonetheless, DDDs are extremely costly in terms of computation time; they do not account for size effects and cannot be readily implemented in standard finite element (FE) software [MONA18].

On the other hand, the approach of *continuum dislocation dynamics (CDD)* describes the microstructure indirectly. The microscopic mechanisms and effects are averaged and transferred to the macroscopic scale by means of microstructural (internal) state variables (MSV). In macroscale FE simulations of plastic deformation, the MSV are calculated at integration points. By means of CDD theory, dislocation density-dependent constitutive models close the gap between the micro-level phenomena and macro-level continuum quantities, such as stress and strain rate [DOMK05]. They can be easily implemented in FE software and are suitable for industrial application [MOTA19]. OROWAN [OROW34] and TAYLOR [TAYL34] first introduced CDD and the relationship between dislocation densities, plastic strain rate and yield stress. WEBSTER [WEBS66], BERGSTRÖM [BERG70] and MECKING and KOCKS [MECK81] developed the model approach by introducing a time-depended change in dislocation density due to multiplication, immobilization and annihilation processes. Subsequently, many models were developed with more than one type of dislocation density (multi-MSV models). A detailed overview can be found in [MOTA19].

ESTRIN ET AL. differentiated dislocations with regard to their arrangement in the dislocation network and introduced a 2-MSV model with few cell and many wall dislocations [ESTR98]. The two dislocation types can interact with each other. In contrast to the model described by ZEHETBAUER [ZEHE93], it is assumed that the volume fractions are not constant and decrease over time due to hardening processes. However, this model developed by ESTRIN only describes the material behavior under quasi-static load conditions ($\dot{\varepsilon}_{\mathrm{pl}}$ = 10^{-4}...10^{-2} s^{-1} [FIEL04]) and was therefore further developed by LEMIALE ET AL. for higher strain rates [LEMI10]. The work of DING ET AL. on milling of aluminum [DING11] and titanium alloys [DING14], laser shock compression of copper [DING12B] and cold rolling of aluminum alloys [DING12A] using the LEMIALE model showed good agreement with the experimental results of dislocation densities.

2.1.4 Intermediate conclusion

Zwischenfazit

In the first chapter of the state of the art, the material class of stainless steels and in particular austenitic stainless steels was discussed. In addition to the increasing demand for stainless steels in established industries such as the automotive and chemical industries, the production of renewable energies will be a great reason for a significant increase in production. The class of metastable austenitic stainless steels in particular offers a wide range of applications due to their excellent work-hardening behavior. In addition to the mechanism of martensite transformation, processes at the microstructural level, especially dislocation generation, movement and accumulation, lead to an increase in strength while maintaining the same ductility. There is a multitude of empirical as well as microstructural material models for the description of plastic deformation as result of dislocation interactions. Thereby, the plastic deformation is significantly influenced by the strain state and the strain rates occurring during cold forming. The microstructure, e.g. twin and shear band formation as well as martensite formation and thus the mechanical material properties can therefore vary over a wide range.

Due to the large number of cold forming processes and thus different load and contact conditions coming from various process parameters, there are no generally valid explanatory models regarding the resulting microstructure and mechanical properties. The cause-and-effect relationships between the forming energies introduced and the material properties must therefore be determined individually for each manufacturing process. Due to the increasing importance of surface finishing in existing manufacturing chains and the focus of this work on the class of mechanical processes, their underlying contact mechanisms and influences on surface integrity as well as fatigue properties will be explained in the following.

2.2 Surface layer treatment

Randzonenbearbeitung

The surface layer of metallic components is the most heavily stressed area of a workpiece [PIPP06]. On the one hand, the workpiece surfaces in contact wear out because of tribological or corrosive stress. On the other hand, micro-crack formation can lead to material fatigue in the surface layer of the component [NICH06]. It has already been proven that by surface layer treatment, both the friction wear can be reduced [KLOC18] and the fatigue strength can be significantly increased [LIEN17].

A large number of surface layer machining processes can be divided into thermal [BRUN01], chemical [COST05], thermochemical [KONG98], physical [HUA06], mechanical [SCHU06] and thermomechanical [WAGN94] processes. However, the surface layer treatments have different effects on the material properties in the surface layer [BRIN14]. For example, a high processing temperature tends to lead to tensile residual stresses, which usually has a negative effect on the service life of the component [TRAU16A]. Mechanical surface layer treatment (MSLT) at room temperature, on the other hand, offers various advantages, which are discussed in more detail below.

2.2.1 Processes for mechanical surface layer improvement

Verfahren zur mechanischen Randzonenverbesserung

When using MSLT, the workpiece surface layer is elasto-plastically deformed by a certain energy input [WOHL00]. A first distinction is made between a static or impact-like energy input. A further distinction is made as to whether a relative movement between the tool and the workpiece exists or not [SCHU06]. An overview of selected direct MSLT as a function of the impact velocity on the workpiece surface and the load application is shown in Figure 2.7a. Laser shock peening [SUND19] and water cavitation peening belong to the group of indirect methods not listed here [BLEI12]. Deep rolling or burnishing and shot peening are the most commonly used MSLT. In contrast, machine hammer peening (MHP) is still a relatively new process.

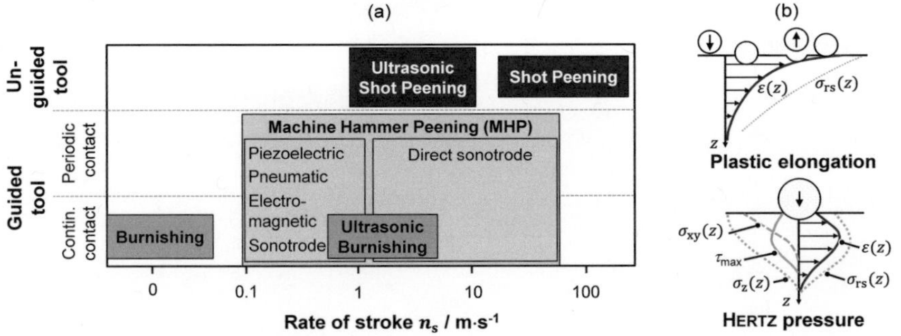

Figure 2.7: Overview of mechanical surface layer treatments according to SCHULZE ET AL. [SCHU16] (a) and schematic illustration of the elementary theories of plastic elongation [WOHL83A] and HERTZ pressure [HERT82]

Überblick über mechanische Randzonenbearbeitungsverfahren nach SCHULZE ET AL. (a) und schematische Darstellung der elementaren Theorien der plastischen Streckung und der HERTZ-Pressung (b)

Characteristic of the non-guided methods, which include shot peening and its derivatives, are stochastic ball impacts. Possible derivatives are the surface mechanical attrition treatment (SMAT) [GROS19], ultrasonic [LI19] and high energy shot peening [LIU01]. Deep rolling (also burnishing and sizing) [MEYE16] and machine hammer peening [TRAU16B] are characterized by continuous or periodic contact due to a guided tool.

Basically, the mechanisms of MSLT can be described with the two elementary theories of plastic elongation [WOHL83A] and HERTZ pressure [HERT82], see Figure 2.7b. The plastic elongation describes that the surface is stretched plastically to a maximum due to many impacts [KLUM19]. The greatest compressive residual stresses often occur directly at the surface and decrease with increasing surface layer depth. HERTZ pressure is characterized by a multi-axial stress state and a shear stress maximum below the surface. The greatest compressive residual stresses occur below the surface. A detailed description of the two elementary mechanisms can be found in [LIEN17]. In the following, the influences on the surface layer resulting from different contact conditions for the above-mentioned MSLT are explained in more detail.

2.2.2 Effects on surface layer

Einfluss auf die Randzone

By means of MSLT, the material properties as a result of the inhomogeneous elasto-plastic deformation in the surface layer can be influenced in a controlled manner. Surface layer properties include, for example, surface roughness, residual stress state and microstructure. THE INTERNATIONAL ACADEMY FOR PRODUCTION ENGINEERING (CIRP) has defined the term *surface integrity* for this purpose. Characteristic surface integrity parameters [CHAT19] are shown in Figure 2.8a.

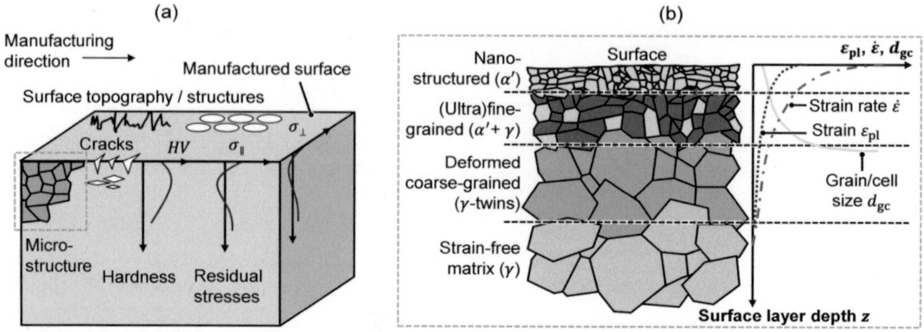

Figure 2.8: Surface integrity according to the CIRP [CHAT19] (a) and influence of surface mechanical attrition treatment (SMAT) on AISI 304 microstructure evolution according to Lu et al. [LU04] and Chen et al. [CHEN11A] (b)
Definition der Oberflächenintegrität gemäß CIRP (a) und Einfluss von SMAT auf die Entwicklung der AISI 304 Mikrostruktur nach Lu et al. und Chen et al. (b)

The greatest work hardening, represented by the material hardness (HV), usually occurs at the workpiece surface [SCHO86]. Further properties associated with the defined surface integrity are the surface roughness or topography, the residual stresses parallel ($\sigma_{||}$) and perpendicular ($\sigma_{\perp}$) to the manufacturing direction and the microstructure. Lu et al. [LU04] and Chen et al. [CHEN11A] provided a detailed examination of the influenced microstructure with different grain sizes in the surface layer as a function of the effective strain conditions after SMAT of AISI 304, see Figure 2.8b. Besides a change in grain or dislocation cell size with varying strain rates and strains, possible α'-martensite transformation or twin formation (orange lines) in the austenitic microstructure (γ) are also shown.

Surface topography

The plastic deformation in MSLT results in roughness changes. Surfaces can be specifically smoothed [BLEI13], roughened [AMAN19] or structured [STEI16]. Depending on the energy applied, material may be thrown up or drawn in at the indentation edges [PINT12]. The MSLT differ with regard to the applied energy and thus the occurring strain rates $\dot{\varepsilon}_{pl}$. For shot peening the strain rates are between $\dot{\varepsilon}_{pl}$ = 10^3...10^5 s^{-1} [SCHU06], for deep rolling they are $\dot{\varepsilon}_{pl}$ = 10^{-1}...10^2 s^{-1} [ALTE03] and for laser shock peening they are $\dot{\varepsilon}_{pl}$ = 10^5...10^{10} s^{-1} [PEYR00]. The influence of the processes on the surface roughness must therefore be evaluated differently. A common objective is the reduction of the surface roughness in order to decrease the notch effect and thus the crack initiation on the workpiece surface.

Materials with a higher initial hardness usually show a lower increase in roughness after shot peening and its derivatives [MART80]. An increase in the peening medium size and hardness as well as the peening pressure (velocity) and thus the kinetic energy also leads to an increased roughness. Holzapfel gives an overview of the roughness change caused by shot peening [HOLZ98], see Figure 2.9a.

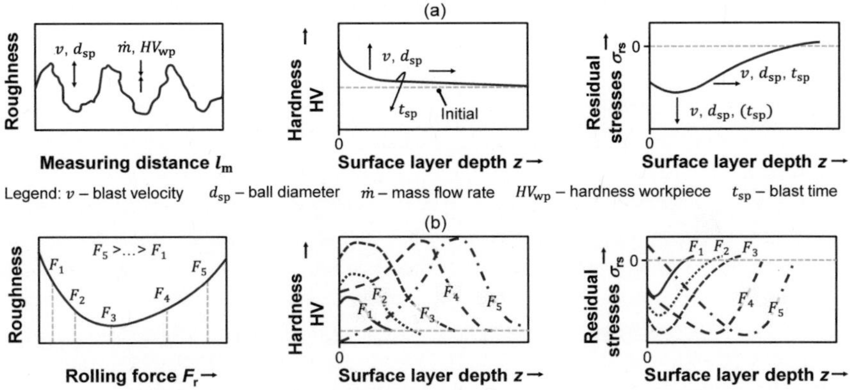

Figure 2.9: Schematic illustration of influence of shot peening process parameters on workpiece roughness [HOLZ98], hardness and residual stresses [VÖHR87] (a) and influence of deep rolling force on workpiece roughness, hardness and residual stresses [FUCH83] (b)

Schematische Darstellung des Einflusses von Kugelstrahlparametern auf die Werkstückrauheit, Härte und Eigenspannungen (a) und Einfluss der Festwalzkraft auf die Werkstückrauheit, Härte und Eigenspannungen (b)

AMANOV ET AL. [AMAN19] and BALUSAMY ET AL. [BALU13] found an increase in surface roughness for shot peening of AISI 304. LESYK ET AL., on the other hand, were able to achieve a reduction in roughness of AISI 304 by using various MSLT [LESY19]. BEN MOUSSA ET AL. also found a reduction in roughness for deep rolling [BEN19] and WANG ET AL. for ultrasonic impact treatment [WANG16]. FUCHSBAUER determined that, depending on the rolling force, deep rolling of a milled surface first leads to a reduction in surface roughness and then to an increase due to damaged areas close to the surface [FUCH83], see Figure 2.9b. However, different initial roughness of the AISI 304 materials investigated must be taken into account.

Residual stresses

The elasto-plastic deformation of the near-surface layers in the tensile area and the elastic deformation of the underlying layers lead to the formation of compressive residual stresses in the surface layer [MADE06]. For this purpose, MADER established a valid layer model for MSLT, with the help of which he explained negative elastic surface-parallel compensating strains as the cause for the resulting compressive residual stresses (macroscopic first type) [MADE06]. Furthermore, residual stresses are generated by volume expansion during the strain-induced martensite transformation [ALTE00]. Due to the multi-phase nature of most materials, phase-specific residual stresses also occur [SCHO91]. Depending on the dominating mechanism, HERTZ pressure or plastic surface elongation, the maximum of the compressive residual stresses is below or directly at the surface [ALHA82]. In shot peening, increasing initial workpiece hardness leads to higher surface residual stresses, lower penetration depths and a pronounced compressive residual stress maximum below the surface, see Figure

2.9a. This is explained by an increasing proportion of HERTZ pressure [WOHL83B]. With increasing peening medium size, peening pressure and velocity and thus resulting higher kinetic energies, the penetration depth and the maximum compressive residual stresses increase.

For deep rolling, the residual stress curve can be described with the help of HERTZ pressure model, whereby the residual stresses transverse to the rolling direction are greater in terms of their magnitude than in the rolling direction [ALTE00], see Figure 2.9b. In most cases, the compressive residual stresses introduced reach at most the stress values of the original yield stress of the material. However, due to strain-induced martensite transformation and high dislocation hardening, compressive residual stresses up to twice the value of the original yield stress can be introduced in metastable austenitic stainless steels [SCHO91]. Nevertheless, an influence of different materials and heat treatment conditions on the residual stress state must be taken into account.

Work hardening

Work hardening is used to describe the increase in strength resulting from the obstruction of dislocation movement by dislocation interactions. For their (qualitative) characterization, strength and hardness parameters are used [HAUK06]. The strength properties are influenced by local lattice distortions due to dislocation processes (residual stresses of the third type) [HABS15]. Depending on the dislocation density in the initial state $\rho_{\mathrm{tot,0}}$, MSLT can be used to achieve pronounced work hardening (low $\rho_{\mathrm{tot,0}}$) up to material softening (high $\rho_{\mathrm{tot,0}}$) because of dislocation annihilation [HOFF01].

The differing local plastic strains and strain rates of the different MSLT have a significant influence on work hardening. High strain rates lead to suppression of thermally activated processes such as dynamic recovery or transverse slip of screw dislocations [HOFF11]. In the case of planar sliding due to low stacking fault energies γ_{SF} in fcc metals, disordered dislocation structures with immobile dislocations were observed with increasing strain rates [MESS14]. In some cases the formation of deformation twins has been detected [ALTE00]. With shot peening, higher blast velocity (kinetic energy) and ball diameter lead to greater effective depth of hardness increase, see Figure 2.9a. In contrast, a higher workpiece hardness and blast time lead to lower hardness values [HOLZ98]. AMANOV ET AL. with shot peening and UNSM [AMAN19], KONGTHEP with deep rolling [KONG19] and GATEY ET AL. with SMAT [GATE13] showed the potential of MSLT to increase surface layer hardness of AISI 304.

Microstructure

MSLT lead to the generation or rearrangement of already existing lattice faults, of which dislocations are the most important. Furthermore, in the case of austenitic stainless steels, twinning and martensite phase transformation occur. The dislocation density is significantly influenced by the plastic strains and the strain rate. The dislocation density is generally increased in soft to medium hard material grades and reduced in hardened steels [SCHO93]. Higher strain rates, such as in the case of shot peening, lead to a pronounced transverse sliding hindrance of the dislocations and thus to randomly distributed dislocation arrangements. Lower strain rates, as in deep rolling, lead to the formation of spatial dislocation cell structures [LU92]. The dislocation structures are formed after the last processing step [NIKI07]. Due to process-related differences, some work on the effect of shot peening and deep rolling on the surface layer microstructure of AISI 304 is considered below.

ALTENBERGER ET AL. showed for deep rolling of an austenitic stainless steel that different layers with strain-induced twins, α'-martensite and high dislocation densities were formed below a nanocrystalline surface layer [ALTE99]. The formation of nanocrystalline, with a grain size d_{gc} = 10...100 nm [TJON04], or sub-microcrystalline, with a grain size d_{gc} < 1 µm [CHEN11B], surface layers is explained by cutting of twins [OLUG20] or the dislocations. Another explanation is the formation of sub-grain boundaries (small angle grain boundaries) and large angle grain boundaries [LU04].

NIKITIN produced a thin nanocrystalline layer in the immediate vicinity of the surface by deep rolling of AISI 304 at room temperature. Below this layer, high-density planar dislocation arrangements, micro twins and deformation-induced martensite were found. He identified an optimum of grain refinement as a function of the process parameters. Furthermore, by means of laser shock compression, he produced a lower dislocation density than with deep rolling, but with a greater hardening depth [NIKI07]. KONGTHEP noticed an increasing proportion of deformation twins due to increased pressure during deep rolling of AISI 304. In addition, a direct transformation of the γ-austenite phase to needle-shaped α'-martensite up to 35% was detected. CHEN ET AL. [CHEN11A] and YE ET AL. [YE14] used SMAT to create a surface layer divided into three areas. The first two areas contained of nano- and ultra-fine grains with a combination of α'-martensite and γ-austenite. In the third area pronounced twin formation was observed. Nanocrystalline layer thicknesses of AISI 304 processed by MSLT from 1 µm [ALTE99] to over 80 µm [AMAN19] were reported.

In general it can be stated that higher strain rates and frequent changes in deformation direction lead to a pronounced grain refinement [KAIB01]. Due to the grain anisotropy in polycrystalline materials, several sliding systems can thus be addressed. Nonetheless, if a critical grain size of a few tenths of a nanometer is reached, an inverse HALL-PETCH mechanism occurs and thus the material softens [TAKE01].

In the upper sections, the effects of selected MSLT on the surface integrity of austenitic stainless steels were discussed. Analysis of the fatigue characteristics of surface layer treated workpieces allows the effectiveness of the surface integrity set by MSLT to be evaluated. A separate consideration of the influencing variables is often difficult due to complex interactions. Nevertheless, in the following different influence factors of surface integrity on fatigue strength of surface layer treated workpieces will be examined.

2.2.3 Influence on fatigue behavior

Einfluss auf das Ermüdungsverhalten

The term *fatigue* describes the material failure under cyclic, i.e. repeated, mechanical, thermal or thermo-mechanical load. Tension-compression, torsion and rotation are the most common types of load [BAYE89]. Based on his test results, WÖHLER recognized the phenomenon that under cyclic loading the material strength is significantly lower than the tensile strength. As a result, he thought about a fatigue-based design of components. The WÖHLER diagram (also known as S-N diagram), which shows the stress amplitude versus the number of load cycles, is still used today for the design of workpieces under cyclic loads [WÖHL58]. A few years later, BAUSCHINGER found out that the accumulation of many very small irreversible plastic microstrains is the cause for fatigue damage [BAUS86]. Based on this, BASQUIN formulated a relationship between stress amplitude and number of cycles to failure for component design in the fatigue strength range [BASQ10]. The fatigue life law was formulated by MANSON [MANS53] and COFFIN [COFF53] and describes a relationship between component service life and plastic strain. It allows for the estimation of material service life independent of temperature influence. Both approaches were combined by MORROW in the following equation. Thereby, the development of the elastic ($\varepsilon_{\mathrm{a,el}}$) and plastic ($\varepsilon_{\mathrm{a,pl}}$) components of the total strain amplitude ($\varepsilon_{\mathrm{a,tot}}$) as a function of twice the number of cycles at break ($2N_{\mathrm{B}}$) is described as follows in Equation (2.5) [MORR65]:

$$\varepsilon_{\mathrm{a,tot}} = \varepsilon_{\mathrm{a,el}} + \varepsilon_{\mathrm{a,pl}} = \frac{\sigma'_{\mathrm{f}}}{E}(2N_{\mathrm{B}})^{b} + \varepsilon'_{\mathrm{f}}(2N_{\mathrm{B}})^{c} \qquad (2.5)$$

with σ'_{f} as the cyclic fatigue strength coefficient (BASQUIN coefficient), E the Young's modulus, b the BASQUIN exponent, $\varepsilon'_{\mathrm{f}}$ the cyclic ductility coefficient (COFFIN-MANSON coefficient) and c the COFFIN-MANSON exponent.

Load controlled fatigue testing

Fatigue can be divided into low cycle fatigue (LCF), high cycle fatigue (HCF) and very high cycle fatigue (VHCF). In the area of LCF, high loading amplitudes occur and the number of cycles to failures is $N < 10^4 \ldots 10^5$. Due to the plastic strains resulting from high load amplitudes, the specimen fails after a few oscillating cycles. The COFFIN-MANSON law is mostly used here. Characteristic for HCF are low loading amplitudes and $N = 10^5 \ldots 10^7$, simplified described by the law of BASQUIN [MUGH15].

An overview of possible fatigue testing methods can be found in [LEE04]. DIN 50100 gives recommendations for testing the fatigue strength of metallic material specimens with constant load amplitude [DIN16]. The already explained WÖHLER test is the best-known test. Cyclic testing of samples on many different load horizons allows complete fatigue strength curves to be established over the three ranges LCF, HCF and VHCF, see Figure 2.10a. The experiments are performed with either constant mean stress σ_{m} or constant load ratio R.

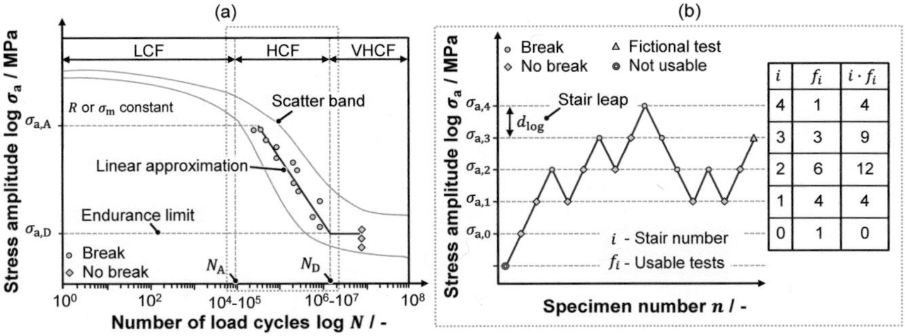

i	f_i	$i \cdot f_i$
4	1	4
3	3	9
2	6	12
1	4	4
0	1	0

Figure 2.10: Schematic progression of a WÖHLER curve [DIN16] (a) and testing arrangement for the stair case method [MÜLL12] (b)
Schematischer Verlauf einer Wöhlerkurve (a) und Versuchsanordnung beim Treppenstufenverfahren (b)

However, this experiments for creating a complete WÖHLER curve requires a large number of specimens [FÜRS19]. Therefore, further fatigue tests are characterized by the fact that they require a much smaller number of specimens but they make only statements in one of the three ranges mentioned above. In the area of fatigue strength, the string-of-pearls and the horizon method are mainly used [DIN16]. In the area of HCF, the stair case method [DIXO48], the probit method [FINN09] and the delimitation method [MAEN77] are possible test methods. An overview of further test methods is given in [RADA07]. The stair case method is most frequently used because it requires little prior knowledge of the expected fatigue strength and gets by with a relatively small number of specimens. The method is characterized by testing in equidistant stair leaps after runs (no break) or breaks on a correspondingly next higher or next lower load horizon, see Figure 2.10b. MÜLLER ET AL. described how a 50% fatigue strength can be calculated from the number of tests on the individual stair leaps [MÜLL12].

Mechanism of fatigue

The five steps *lattice defect* (hardening, softening), *slip tears* (strain localization), *occurrence of transcrystalline micro cracks*, *growth of micro cracks* and *fracture* describe the fatigue mechanism [MUGH15]. Cracks often occur first on the surface at grain boundaries or inclusions [COUR05]. During forming, the movement of slip bands is hindered by dislocations and stress concentrations occur at the crack tips. If a critical stress is exceeded, crack growth occurs is described with the PARIS law [BRIN08].

For materials, which harden cyclically, the hardening is usually completed after 5% to 10% of the service life [SKEL87]. The cyclic deformation properties depend very much on the test temperature and the strain rate [KANA88]. The formation of martensite in case of metastable austenitic stainless steels depends largely on the strain amplitude and temperature [BAUD77]. In his work, NIKITIN described the influence on dislocation activities during cyclic loading, which also results from this relationship [NIKI07]. LUKAS ET AL. explained the dislocation structures as a function of the number of cycles to failure and stacking fault energy for fcc metals, see Figure 2.11a [KLES92].

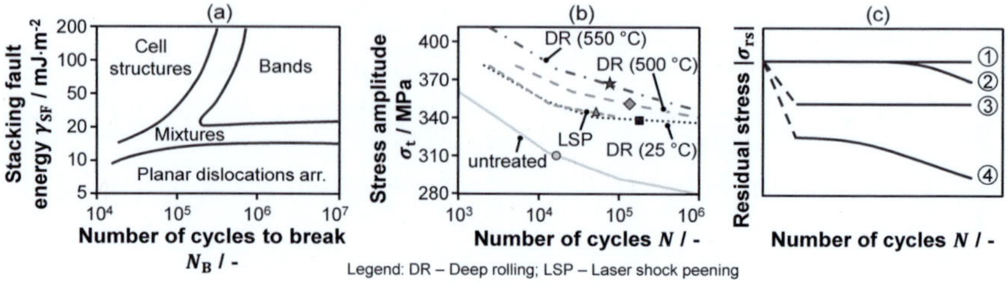

Figure 2.11: Dislocation structures as a function of number of load cycles for fcc metals [KLES92] (a), WÖHLER graphs of treated AISI 304 [NIKI07] (b) and types of cyclic residual stress relief according to MACHERAUCH and WOHLFAHRT [MACH85] (c)
Versetzungsstrukturen in Abhängigkeit von der Lastspielzahl für kfz-Metalle (a), WÖHLER-Kurven von behandeltem AISI 304 (b) und Varianten von zyklischem Eigenspannungsabbau nach MACHERAUCH und WOHLFAHRT (c)

Fatigue of surface layer treated materials

In stress controlled testing, an increased proportion of hardened material leads to a significant reduction in the plastic strain amplitude for the entire component, which in turn leads to smaller microstructural changes during cyclic deformation. The higher the hardened surface layer percentage, the higher the cyclic yield strength [ALTE00]. NIKITIN analyzed the influence of different MSLT processes on the fatigue strength of AISI 304. He showed that by means of MSLT the fatigue strength of AISI 304 can be increased significantly, see Figure 2.11b [NIKI07].

While compressive residual stresses have a positive effect on crack propagation behavior [SCHO86], work hardening has a positive influence on crack formation. Furthermore, work hardening reduces plastic deformations under cyclic loading [SCHU06]. Pronounced work hardening also counteracts the reduction in residual compressive stress [TOTT02]. With the cyclic loading of AISI 304, the amount of residual compressive stresses introduced has no significant influence on the fatigue life of the material, but the reduction behavior of the residual compressive stresses does [LAMP96]. The reduction of residual stresses under cyclic loading depends on the material strength [VAND81]. Macro residual stresses are quickly reduced in soft steels and therefore have little influence on fatigue strength. In medium-strength steels, macro residual stresses only become irrelevant at high load amplitudes in the fatigue strength range. Due to their long-term stability, macro residual stresses play a decisive role in high-strength steels. Work hardening or micro residual stresses due to dislocation interactions have a major influence on the fatigue strength of soft steels. The reduction of residual stress under cyclic loading is determined by the plastic strain amplitudes that occur and can be divided into the following four cases (see Figure 2.11c):

1 - Residual stresses remain stable because neither the quasistatic nor the cyclic yield strength is exceeded.

2 - Only with a large number of load cycles one can observe a reduction of residual stress, because only the cyclic yield strength is exceeded.

3 - Only the quasistatic yield strength is exceeded, no further residual stress reduction after the first cycle.

4 - The cyclical and quasi-static yield strength is exceeded, pronounced reduction of residual stresses in the first load cycle and continuous further reduction of residual stresses

Since a high surface roughness favors crack initiation due to the notch effect, the smoothest possible surfaces are to be selected for high fatigue strength. Roughness has little influence on crack propagation [KRUP07].

2.2.4 Intermediate conclusion

Zwischenfazit

The contact mechanisms of the mechanical surface layer treatments (MSLT) can be described by means of the elementary theories according to HERTZ and plastic elongation. The most commonly used MSLT are shot peening, its derivatives and deep rolling. The methods differ with regard to the load spectra and forming energy occurring during processing. These, in turn, have a significant influence on the strain states and strain rates appearing in the workpiece surface layer. The resulting properties of the surface integrity can vary accordingly in a wide range. In general, MSLT can be used to introduce compressive residual stresses and work hardening in the workpiece surface layer and to achieve phase transformation and microstructural sub-grain formation. Furthermore, depending on the process parameter selection, surfaces can be smoothed or roughened. The fatigue strength of surface layer treated workpieces is thereby mostly increased.

Machine hammer peening (MHP) is a new type of MSLT, which generates deterministic, high-frequency spherical impacts on the workpiece surface. In principle, the MHP combines the advantages of shot peening and deep rolling. However, the process kinetics resulting from the underlying MHP working principle and thus the strain states caused in the workpiece surface layer are still largely unknown. Therefore, the MHP will be discussed in more detail below.

2.3 Machine hammer peening (MHP)

Maschinelles Obeflächenhämmern (MOH)

Machine hammer peening (MHP) is an incremental forming process according to DIN 8580 for the surface layer treatment of metallic materials [DIN03]. In general, MHP is used for post-weld treatment [SMID09], to increase the service life of ball and slide bearings, cutting blades and cam followers [AMAN11] and as a polish substitute in automotive tool and mold making [MANN17]. The VDI guideline 3416 gives recommendations regarding the technological terminology, e.g. the desired effects smoothing, structuring, hardening as well as the introduction of compressive residual stresses [VDI18]. According to BLEICHER ET AL. the MHP is to be assigned to the direct, deterministic mechanic surface layer treatment (MSLT) on the basis of a defined feed system and the tool directly involved in the forming process [BLEI12]. Due to its process kinematics, the MHP combines the advantages of conventional MSLT such as deep rolling or shot peening [TRAU15A]. Furthermore, the MHP can be applied in a workpiece clamping to a previous machining process. Thus, the set machine reference point can be kept without additional programming [TRAU16A].

2.3.1 Technological process variations

Technologische Prozessvarianten

A characteristic feature of all MHP processes is the linear axial hammer head movement in the direction of the workpiece. Thereby, the necessary forming work is provided by the kinetic energy of the moving mass [VDI18]. KRALL divided the MHP process variants into systems with periodic contact and generated individual indentations (e.g. electrodynamic, piezoelectric) and systems with continuous contact and continuous paths (e.g. pneumatic) [KRAL19]. Periodic MHP tool systems offer the possibility of smoothing or structuring workpiece surfaces [STEI13]. A further possibility to differentiate between MHP systems is to consider their underlying physical principle of operation. The MHP systems work according to the electrodynamic [LÖCK07], pneumatic [WIED10], piezoelectric [KLAM10], mechanical [OSTE16] and the electromagnetic [WEIL19] operating principle. A detailed presentation and explanation of the different MHP operating principles with regard to hammering frequency and impact energy range as well as the development of MHP patents can be found in [KRAL19]. Depending on the technology used, different process variants have various effects on the properties of the surface layer of a workpiece [WIED11A].

By initiating the relative movement between the MHP tool and the workpiece, complete workpiece surfaces can be machined incrementally [RODR14]. Depending on the type of relative movement, plane, cylindrical and freeform surfaces can be machined by means of MHP. Currently, MHP tools and control units in combination with industrial robots or machining centers are used [SCHU16]. When guiding the MHP tool through a multi-axis machining center, the motor spindle does not rotate because of pneumatic and electric connections. Thereby, it could be shown that false brinelling wear on the bearing rings occur at hammering frequencies f = 100 Hz [MANN18A]. Disadvantages of an industrial robot as MHP guide system are lower stiffness and resulting vibrations, which can influence the processing result [KRAL16]. KRALL [KRAL19] and HACINI ET AL. [HACI09] carried out investigations with different types of industrial robot.

Analogous to other MSLT, the plastic deformation of the surface layer during MHP causes surface smoothing [SCHE13], induction of work hardening [ADJA12] and compressive residual stresses [MANN18B] depending on the selected MHP process parameters. Furthermore, defined surface structures can be generated by means of MHP, which have friction-reducing properties due to elastostatic and -hydrodynamic effects [KLOC14]. It has also been demonstrated that MHP can be used to adjust grain refinement in the surface layer [CHEN13] and hammering hard material particles into metallic surfaces to improve frictional wear resistance [BLEI13]. For quenched and tempered steel AISI 4140 (42CrMo4, 1.7225), LIENERT showed that by means of piezoelectric MHP the fatigue strength could be significantly improved compared to deep rolled and shot peened specimen [LIEN17].

In this work, the electrodynamic MHP principle is used. Therefore, it will be discussed in more detail in the following.

2.3.2 Electrodynamic operating principle

Elektrodynamisches Wirkprinzip

The operating principle of electrodynamic MHP is based on the local plastic deformation of the surface layer by high-frequency impacts (up to 500 Hz) of a mostly spherical hammer head [TRAU13]. Due to the large number of adjustable process parameters, the electrodynamic MHP is characterized by high process flexibility and high impact energy [TRAU15B]. The adjustable process parameters of the electrodynamic MHP are the hammer head diameter d, the contact energy E_c, the hammering frequency f, the stroke h, the feed rate v, the stepover distance s, the path orientation angle α_{mhp} and the resulting impact angle β_i [VDI18]. The first three parameters are defined directly via the MHP apparatus. The remaining MHP process parameters result from the relative movement between the tool and the workpiece [HABE15]. An overview of selected MHP process parameters is shown in Figure 2.12.

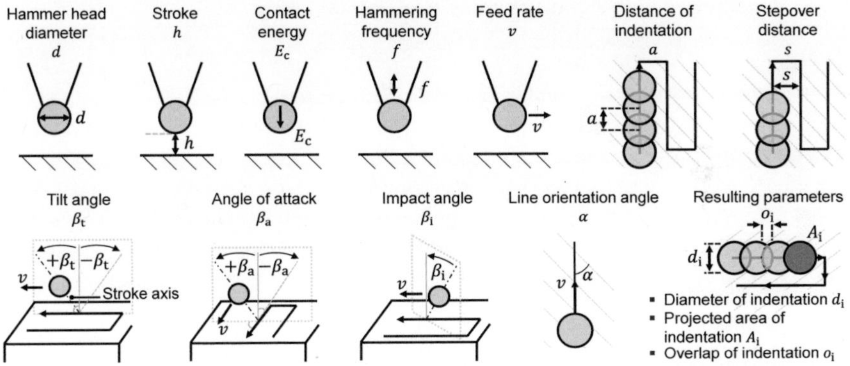

Figure 2.12: Overview of selected MHP process parameters according to VDI 3416 [VDI18]
Übersicht ausgewählter MOH-Prozessparameter nach VDI 3416

The operating principle is based on an electromechanical coupling [GROC12]. By applying an alternating voltage to a copper inductor, which is located in the magnetic field of a permanent magnet, an axially directed magnetic force (LORENTZ force) is generated [LÖCK12]. This force accelerates the plunger with screwed-on hammer head located in the center of the inductor until it hits the workpiece. Thereby the kinetic energy is converted into forming work. The rebound energy resulting from the elasto-plastic forming process is used to move the plunger back to the starting position in the direction of the inner mechanical stop and thus to maintain the process [HABE15]. In Figure 2.13a, a commercial electrodynamic hammering system type 2002 of ACCURAPULS

GmbH is attached to an industrial robot ABB IRB6660-205/1.9. Using CAD-CAM software, path and machining strategies can be developed to protect the workpiece edges during machining. Figure 2.13b shows the schematic structure of the MHP tool.

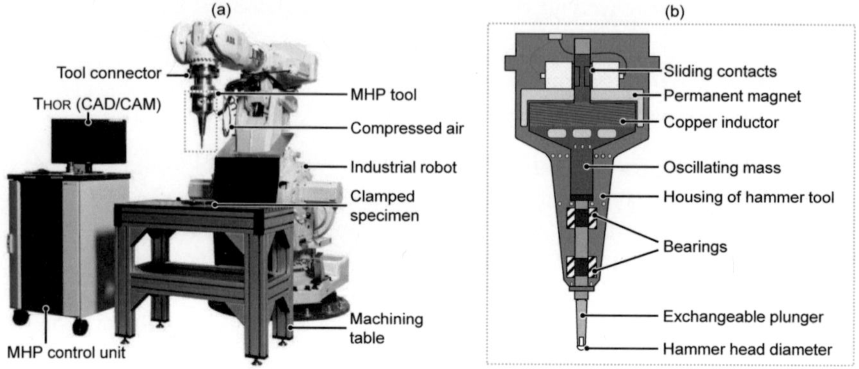

Figure 2.13: Connection of an electrodynamic MHP system on an industrial robot (a) and schematic of the electrodynamic operating principle (b)
Adaption eines elektrodynamischen MOH-Systems an einen Industrieroboter (a) und Schema des elektrodynamischen Wirkprinzips (b)

2.3.3 Kinematics, contact mechanics and impact analysis

Kinematik, Kontaktmechanik und Einschlaganalyse

Electrodynamic MHP is a multiphysical production system in which electrical energy is converted into mechanical (kinetic) energy. The contact energy finally results from the kinetic energy of the accelerated mass in the hammer head, taking into account loss terms such as heat and friction. Thus, the knowledge of the kinetic energy is essential since it has a decisive influence on the contact energy and the surface layer properties. For this reason, several models have already been established to describe the process kinematics of electrodynamic MHP.

Process kinematics

Wied firstly modeled the process kinematics using Newton's second law of translational motion. He discovered proportional relationships between the electric current and hammering force. Thereby, he only considered the Lorentz force and neglected losses due to friction, bearings or motional electromotive forces [WIED11A]. Bleicher et al. extended Wied's model regarding the equations of motion. They combined the model extension with the Zerilli-Armstrong material model to simulate the plunger velocity, displacement and acceleration. They also neglected motional electromotive forces [BLEI12]. Habersohn extended the model of Bleicher et al. to include workpiece dynamics. In addition, he divided the hammer into three coupled masses in order to consider the hammer head and the drive unit separately [HABE15]. He used the concept of the coefficient of restitution e, which he assumed to be a constant quantity. Trauth used a similar approach to Bleicher et al. by dividing the MHP process into subsystems and additionally considering motional electromotive forces. Furthermore, he chose the half space theory according to POPOV [POPO09] and the flow stress formulation according to HOLLOMON [HOLL45] to describe the elasto-plastic domain [TRAU16A]. However, the models presented have never considered all aspects of electrodynamic MHP. While WIED and BLEICHER ET AL. disregarded motional electromotive forces, HABERSOHN neglected the velocity dependence of the coefficient of restitution and TRAUTH neglected the frequency and power dependence of the coil inductance. The deviations in the calculated contact energy also influence the consideration of the resulting surface integrity (elasto-plastic contact conditions).

Contact mechanics

Due to the feed rate during MHP, there is a tangential contact problem. However, the shear stresses resulting from the surface roughness and material properties of the two contacting bodies hammer head and workpiece are negligible due to the low contact energies and contact time [GHAE17]. Therefore, adhesive and frictional forces [POPO18] are not considered in simplified MHP models.

Because of the high-frequency process characteristics and the simultaneous objective of sufficient impact coverage on the surface, it is not possible with MHP to analyze the influence of individual impacts and the associated energy input on surface integrity. For this reason, WIED developed a drop tower test rig to create a load analogy to the MHP. He investigated the separated influence of mass m, hammer head diameter d and impact velocity v_c on the indentation geometry and energy balances of different materials [WIED11A]. Using supplementary FE simulations of the drop tower tests, WIED ET AL. were able to derive fundamental cause-and-effect relationships between influencing parameters and resulting impact geometries [WIED11B].

Based on the work of WIED [WIED11A] and STEITZ [STEI16], STICHT ET AL. derived an analytical relationship for description of an energy threshold for smoothing of rough surfaces. The model validation was realized by means of experimental drop tower tests [STIC16]. They also used findings from the work of HERTZ [HERT82], JOHNSON [JOHN85], TABOR [TABO00] as well as KIMURA and CHILDS [KIMU99]. HERTZ described the contact of ideally elastic and smooth bodies and neglected plastic deformation [HERT82]. JOHNSON developed an analytical model for dynamic contact situations of smooth bodies considering plastic deformation [JOHN85]. TABOR derived an analytical model for static contact of rough surfaces respecting plastic deformation [TABO00]. KIMURA and CHILDS analyzed dynamic contact problems of rough surfaces by means of numerical approaches [KIMU99]. LECHNER presented a further possibility to investigate the influence of single impacts on MHP in his work. By selecting high machine feed rates, individual indentations were created and their geometry determined as a function of the plunger stroke and the hammer head diameter [LECH14].

However, due to the impact velocities of the MHP, which were still unknown at that time, and the resulting contact energy ranges, a direct transfer of results from the drop tower test rig to the MHP was only possible to a limited extent. The knowledge of the impact velocity and the resulting contact energies, would enable the targeted adjustment of comparable energetic states in analogy test stands. Furthermore, the above-mentioned analyses were limited to the resulting indentation geometry (diameter and depth) and energy balance. However, the strain states prevailing in the materials and their influences on the activated mechanisms of surface integrity and especially the resulting microstructure were not considered. Such a consideration is essential, especially for the material class of metastable austenitic stainless steels and can be done by impact analysis.

Impact analysis

The contact energy resulting from the impact velocity in electrodynamic MHP significantly influences the strain state and thus the plastic deformation area of the workpiece surface layer. Especially the occurring strain rates have a significant influence on the microstructure characteristics. BOYCE ET AL. conducted FE impact simulations for analyzing the residual stress state of a titanium alloys using contact energies E_c = 2.7…6.1 J. The simulated compressive residual stresses were in good agreement with experimental XRD measurements. One region at the surface directly outside the crater rim showed tensile hoop stresses. In the plastic region, the highest tensile stresses were found, see Figure 2.14a. The different areas were defined by JOHNSON [JOHN85]. Compressive residual stresses appeared in a large zone directly beneath the indentation crater [BOYC01].

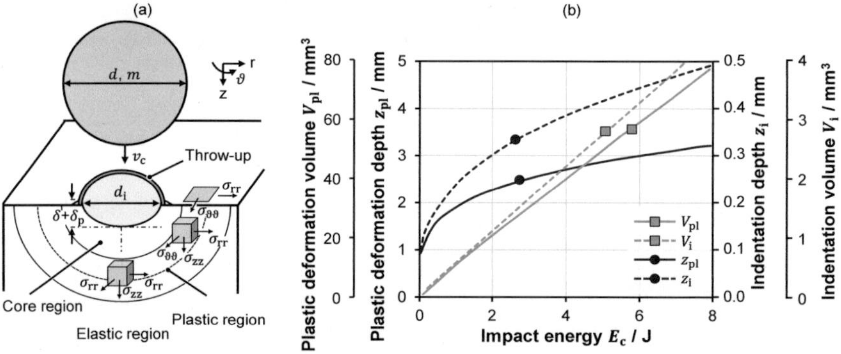

Figure 2.14: Schematic representation of the stress conditions when a sphere hits a flat surface [BOYC01] (a), development of the indentation geometry and the plastically influenced surface layer as a function of the impact energy [KANO12] (b)
Schematische Darstellung der Spannungszustände beim Einschlag einer Kugel auf eine flache Oberfläche (a), Entwicklung der Eindruckgeometrie und der plastisch beeinflussten Randzone in Abhängigkeit von der Einschlagenergie (b)

KANOU ET AL. investigated the influence of different impact velocities v_c = 25…50 m/s on the depth and volume of indentations and the plastically deformed region, see Figure 2.14b. It could be shown that the indentation volume and plastic deformation volume were almost linearly proportional to the impact energy E_c [KANO12]. For visualization of the deformed layer, they used the FRY etching method, which was for impact analysis firstly used by KAI ET AL. [KAI08]. Later, KORZYNSKI developed a dynamic model for predicting the maximum depth of surface roughness dimples after impact treatment [KORZ19]. WIED discovered that the indentation diameter increases with increasing contact energy and increasing ball diameter for various materials [WIED11A].

Furthermore, CHAN ET AL. found out on AISI 304 that the impact velocity is proportional to the strain rate of materials. Thereby, a higher strain rate is believed to lead to more desirable nanostructures in the material and that is more efficient in generating these nanostructures, e.g. nanotwins [CHAN10]. CHEN ET AL. explained that the deformation mechanisms in austenitic stainless steel are significantly influenced by the strain rates [CHEN11B].

2.3.4 Intermediate conclusion

Zwischenfazit

Machine hammer peening is an innovative high-frequency mechanical finishing process that is used to adjust the surface integrity and improve the fatigue strength of highly stressed workpieces. While there is already a certain empirical knowledge about the effect of individual process parameters on the resulting surface layer properties, the underlying mechanisms are still largely unknown. Especially the strain states and strain rates in the workpiece surface layer resulting from the contact energy during MHP play an important role regarding the effects on the surface integrity and thus the fatigue properties. However, the contact energies resulting from the impact velocities in robot-based MHP are still unknown. Correlations with the strains and strain rates occurring and microstructural mechanisms, especially with metastable materials such as austenitic stainless steels, could not be developed so far. By means of the precise knowledge of the contact conditions of MHP and the mapping of the loading conditions of a single impact, e.g. in a drop tower test rig, the MHP contact mechanisms can be investigated in a targeted manner.

2.4 Economic and scientific problem statement

Wirtschaftliche und wissenschaftliche Problemstellung

At the beginning, the state of the art showed the economic relevance of the material group of austenitic stainless steels. Due to the changing social, political and technological boundary conditions, e.g. in the field of renewable energy production, a strong increase in the stainless steel production volume is predicted. The growing market for the production of renewable energies will inevitably lead to new applications and thus to new sales markets for stainless steel products. By 2027, the stainless steel market in China, which leads the world in stainless steel production volume with more than 50%, is expected to almost double. Thereby, the metastable austenitic stainless steel grades are responsible for three-quarters of the world's stainless steel production and offer a broad application potential due to their material properties. These are, above all, the excellent work hardening and transformation properties, whereby the workpiece properties can be adjusted within wide limits and specifically to the respective application. The microstructural mechanisms of dislocation interaction and martensitic phase transformation can be utilized and property-graded components can be manufactured by means of targeted forming. The wide range of current and future areas of application for austenitic stainless steel components poses special challenges for production technology and especially for finishing processes.

This result in the following **economic problem statement**: *Future new fields of application for stainless steel components will influence entire production chains. In particular, the use of mechanical finishing processes promises the production of property-adapted components at low investment. Nonetheless, it is still unclear how these mechanical finishing processes can optimize cost efficiency while simultaneously increasing profitability.*

The established finishing processes in the field of mechanical surface layer treatment include shot peening and deep rolling. Extensive basic and user knowledge is already available for both processes. The innovative process of machine hammer peening (MHP) contrasts this. This process combines the advantages of shot peening and deep rolling due to its kinematics. In addition to the introduction of high and far-reaching compressive residual stresses and work hardening, surfaces can be smoothed or specifically structured and the microstructure of metastable materials can be influenced. The improvement in component fatigue strength resulting from this modified surface integrity predestines the MHP process for the finishing of highly stressed components from austenitic stainless steels. However, the insufficient understanding of the process kinetics and the underlying mechanisms of action during forming currently still hamper a broad industrial application of MHP. Furthermore, it is still largely unclear how the lower system rigidity in highly flexible robot-guided MHP compared to MHP processing on machining centers affects the machining result. While mostly empirical cause-and-effect relationships between the process parameters and surface layer properties have been determined, the understanding of the contact mechanisms resulting from the process kinetics is still largely lacking. Especially the strain state essential for surface integrity and the prevailing strain rates are unknown. This follows from the still mostly unknown contact energies in MHP. A corresponding correlation with the surface integrity, microstructural changes and fatigue properties resulting from robot-based MHP machining of metastable austenitic stainless steels has also not yet been established.

This result in the following two **scientific problem statements**: *Firstly, the cause-and-effect relationship between the contact energies resulting from the robot-based MHP impact velocities and the strain conditions in the surface layer of metastable austenitic stainless steels are unknown. Secondly, the mechanisms of action in the hammered workpiece surface layer resulting from the strain states as well as the surface integrity and specifically the microstructural properties are unknown and therefore knowledge-based MHP machining to increase the workpiece fatigue strength is not yet possible.*

3 Objective and research methodology

Zielsetzung und Forschungsmethodik

The following **economic problem statement** was derived from the current state of the art in science and industry presented in the previous chapter: *Future new fields of application for stainless steel components will influence entire production chains. In particular, the use of mechanical finishing processes promises the production of property-adapted components at low investment. Nonetheless, it is still unclear how these mechanical finishing processes can optimize cost efficiency while simultaneously increasing profitability.*

As a possible solution, the use of the incremental mechanical surface layer treatment method, machine hammer peening (MHP), was derived. The extensive review of the scientific literature has shown that despite intensive research efforts, there are still significant technological and scientific knowledge deficits in the field of MHP. Especially the lack of understanding on the machining influence of an industrial robot in MHP, the contact energies coming from the process kinetics, the resulting strain states in workpieces and thus the surface integrity and finally the fatigue properties lead to a limited industrial application of this technology. For this reason, the following two **scientific problem statements** were formulated: *Firstly, the cause-and-effect relationship between the contact energies resulting from the robot-based MHP impact velocities and the strain conditions in the surface layer of metastable austenitic stainless steels are unknown. Secondly, the mechanisms of action in the hammered workpiece surface layer resulting from the strain states as well as the surface integrity and specifically the microstructural properties are unknown and therefore knowledge-based MHP machining to increase the workpiece fatigue strength is not yet possible.*

Taking into account the economic and scientific problems, the following **research objective** is formulated:

Knowledge-based explanatory model of the cause-and-effect relationships between the contact energies resulting from different process states in robot-based electrodynamic machine hammer peening and the workpiece surface layer of austenitic stainless steels as well as their effects on fatigue properties.

The formulated research objective is to be achieved by investigations based on the following **research hypothesis**:

By means of the defined energy input during robot-based electrodynamic machine hammer peening of austenitic stainless steel the surface integrity of the workpiece can specifically be adjusted and thus their fatigue strength can be improved.

To validate the research hypothesis and to achieve the research objective, the following **four research questions**, which also represent the structure of this doctoral thesis (Figure 3.1), must be answered:

1. How do the process parameters of robot-guided electrodynamic MHP influence the process kinetics and thus the resulting contact energy and how can the multiphysical system behavior be modeled to predict the process kinetics?
2. How do defined contact energies of electrodynamic MHP affect the contact conditions and the geometry of individual impacts and how do these influence the surface integrity of an austenitic stainless steel?
3. How do defined contact energies influence the surface integrity of austenitic stainless steels in the electrodynamic MHP?
4. What is the relationship between a knowledge-based setting of the surface integrity in electrodynamic MHP and the resulting fatigue strength?

In Chapter 4, the **research question 1** is answered. In order to understand the system-related influences of robot-guided electrodynamic MHP on the later machining result, both the robot behavior and the kinetics of the MHP system are investigated. Especially the first inline recording of the impact velocities of the hammer plunger and the hammering forces allows an extended system description depending on the contact energies. Based on this, numerical modelling of the MHP system behavior for the estimation of the process kinetics and contact energies is carried out at the end of Chapter 4 to answer the second part of research question 1.

Once the MHP system behavior is known, the effects of the energy input resulting from different process states on the geometric characteristics of single impacts and their surface integrity will be investigated by means of **research question 2** in Chapter 5. By using an analogy test bench, indentation geometries are analyzed depending on the contact energies. The elasto-plastic stress and strain state in the surface layer allows conclusions to be drawn about the plastic deformation zone as well as the strain rates decisive for the resulting surface layer properties.

Research question 3 in Chapter 6 deals with the topic of the two-dimensional surface layer treatment by means of electrodynamic MHP. Experimental and numerical approaches are used to investigate the plastomechanical stress states and surface layer morphologies resulting from different contact energies. By using a dislocation density-based material model, knowledge of the development of dislocation density in MHP allows further conclusions to be drawn about the mechanical material properties. Finally, the findings are synthesized in a design model, with the help of which an MHP parameter range is proposed that accounts for improved surface layer properties.

Based on the results of the preceding research questions, specific surface layer influences on fatigue strength will be investigated in Chapter 7 by means of experimental tests on a rotating bending test bench. **Research question 4** is concerned with the analysis of the mechanisms that have a significant influence on the fatigue strength of austenitic stainless steels. The stated research hypothesis will be validated through the synthesis of all results in an explanatory model.

The research methodology underlying this doctoral thesis contains both inductive and deductive elements. Empirical studies form the basis of each chapter. By means of experimental analyses, physical phenomena are observed and supported by numerical investigations, and explanatory approaches are derived from a scientific perspective. In addition, this work contains derivations of analytical elements, which describe and explain the physical principles of a multiphysical manufacturing process.

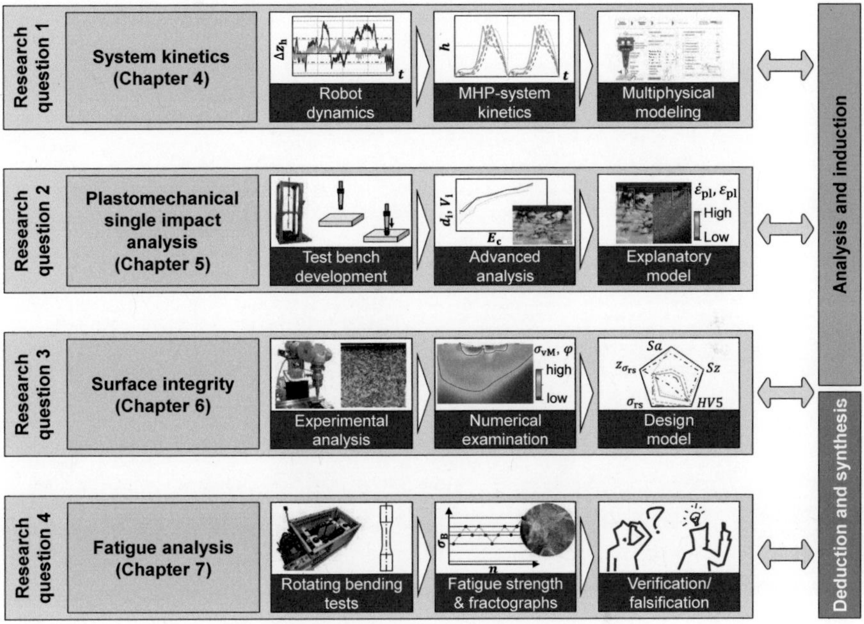

Figure 3.1: Research methodology and approach within this work
Forschungsmethodik sowie Lösungsweg innerhalb dieser Arbeit

4 Characterization and analysis of robot-based MHP kinetics

Charakterisierung und Analyse der roboterbasierten MOH-Kinetik

The review of the state of the art in Chapter 2 has shown that machine hammer peening (MHP) is a mechanical manufacturing process, which is predominantly used in tool and mold making. In MHP, kinetic energy is converted into forming work, plasticizing the workpiece surface layer and specifically influencing the surface integrity. The MHP is a deterministic process, which must be precisely positioned by a carrier system and guided during machining. Therefore, the use of MHP is limited in most cases to machining centers with high rigidity. Thereby, system-related deviations in the MHP processing result can be almost completely excluded. On the other hand, robots as carrier systems offer the possibility of high flexibility with regard to process control and workpiece geometry. Currently, however, MHP processing using industrial robots has so far only been investigated by KRALL ET AL. [KRAL19]. Due to the usually lower and deviating rigidity of industrial robots – even within the robot classes – compared to machining centers, these effects on the MHP processing result must be investigated. Furthermore, there are knowledge deficits in the area of process kinetics of MHP processes. Because of the MHP process characteristics, according to STICHT ET AL. [STIC16], an energetic consideration is purposeful to describe the process behavior and requires knowledge of the MHP contact energies. The knowledge of the energy input into the workpiece allows the derivation of cause-and-effect relationships with the resulting surface integrity. Due to the lack of sensors on the hammer tool, the hammering forces and plunger strokes as well as the resulting impact velocities and contact energies cannot be determined yet.

For the reasons mentioned above, the dynamic behavior of an industrial robot as MHP carrier system and its influence on deterministic MHP process control is examined in Chapter 4.1. In Chapter 4.2, an electrodynamic MHP system is further developed by means of sensors in order to enable inline measurement of relevant electrical (current, voltage) and mechanical (plunger stroke, hammering force) values. The experimentally determined values are presented and discussed in Chapter 4.3. The findings are then synthesized and extended in a numerical model for a realistic representation of the electrodynamic MHP process kinetics in Chapter 4.4. Chapter 4.5 summarizes the results and answers the research question **"How do the process parameters of robot-guided electrodynamic MHP influence the process kinetics and thus the resulting contact energy and how can the multiphysical system behavior be modelled to predict the process kinetics?"**

4.1 Determination of robot dynamics

Bestimmung der Roboterdynamik

Due to the lower stiffness of an industrial robot compared to a machining center, the robot dynamics and its influence on the hammering result were investigated in this chapter. Acceleration studies were used to analyze in which areas the robot has a constant velocity, thus ensuring reproducible contact conditions for MHP. The oscillation of the robot axis with the hammer tool was investigated to evaluate the machining influence on the workpiece topography and the contact energy. Furthermore, the elastic deformation of the oscillating mass during MHP was analyzed. This procedure enables a qualitative evaluation of the robot influences on the MHP processing result.

4.1.1 Analysis of robot acceleration

Analyse der Roboterbeschleunigung

For the analysis of the robot dynamics, the electrodynamic MHP system type 2002 from ACCURAPULS was adapted to an industrial robot IRB6660-205/1.9 from ABB, see Figure 4.1a. This six-axis robot has a working range l_{ra} = 1.9 m and a handling capacity m_{r} = 205 kg. The total mass of the hammer tool, which is attached to the robot flange via an aluminum adapter, was determined to be $m_{\mathrm{tot}} \approx$ 12.25 kg.

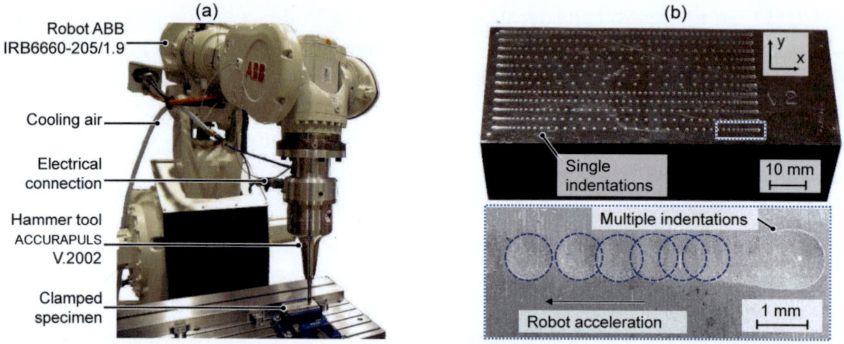

Figure 4.1: Experimental setup for analysis of robot acceleration (a) and workpiece with hammered paths for single indentation evaluation (b)
Experimenteller Aufbau zur Analyse der Roboterbeschleunigung (a) und Werkstück mit gehämmerten Bahnen zur Auswertung einzelner Eindrücke (b)

Firstly, the acceleration profiles of the robot were determined. By knowing the acceleration range in which the robot achieves a constant feed rate, the process control for the MHP tests can be specifically adapted. This enables the setting of reproducible and comparable processing states of the workpiece in further MHP tests. For the analysis of the acceleration profiles of the robot with adapted hammer tool, test series with feed rates v_{r} = 120...340 mm·s^{-1} were carried out. The selected feed rates were higher than for industrial MHP machining. This allows the application of individual, non-overlapping indentations and thus enables an evaluation of the acceleration ranges of the robot. The influence of increased friction was neglected since the results from LECHNER already showed that despite the high feed rates, circular and not elliptical indentations are formed [LECH14]. Additionally, the hammering frequencies f = {80; 120; 180} Hz were varied. These are standard hammering frequencies used in the industry. The hammer head diameter d = 6 mm, the electrical power stage P_{el} = 70% and the stroke h = 0.8 mm were kept constant. The value of the electrical power stage corresponds to the maximum value recommended by the manufacturer in continuous MHP operation. By evaluating the distances between the resulting indentation diameters d_{i}, see Figure 4.1b, the feed rate of the robot v_{r} was calculated according to Equation (4.1):

$$v_{\mathrm{r}} = f \cdot (x_{\mathrm{d_{i+1}}} - x_{\mathrm{d_i}}) \quad \textbf{(4.1)}$$

where $x_{\mathrm{d_i}}$ is the position of one indentation along the feed axis. The indentations were evaluated using a Marvision MM 320 microscope from MAHR GmbH. The indentation diameter was determined as d_{i} = 0.930...1.065 mm. Due to partly blurred indentation edges, a small measurement inaccuracy could not be excluded. The tests were carried out along a path length l_{x} = 70 mm for a sufficient acceleration distance. The results of the acceleration profiles for the hammering frequencies f = 80 Hz (a), f = 120 Hz (b) and f = 180 Hz (c) are shown in Figure 4.2.

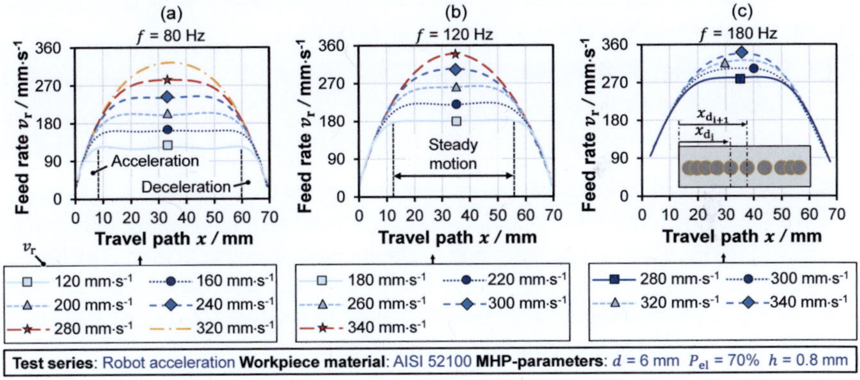

Figure 4.2: Evaluation of robot acceleration by different feed rates v_{r} for the hammering frequencies f = 80 Hz (a), f = 120 Hz (b) and f = 180 Hz (c)
Auswertung der Roboterbeschleunigung bei verschiedenen Vorschubgeschwindigkeiten v_{r} für die Hämmerfrequenzen f = 80 Hz (a), f = 120 Hz (b) und f = 180 Hz (c)

The acceleration profiles of the robot has been divided into three areas: acceleration, steady motion and deceleration. Depending on the selected feed rates, the average length with constant feed rate is reduced over the measured distance of l_{r} = 70 mm. It could be determined that independently of the selected hammering frequency f and the set feed rates, the acceleration ($a_{\mathrm{r}} > 0$) and deceleration ($a_{\mathrm{r}} < 0$) can be described with the following Equations (4.2) and (4.3) for the feed rate v_{r}:

$$a_r > 0: \quad v_r(x) = 7.14 \cdot x + 100.9 \quad \textbf{(4.2)}$$

$$a_r < 0: \quad v_r(x) = -6.22 \cdot x + 549.5 \quad \textbf{(4.3)}$$

In industrial MHP operation, work is usually carried out with a hammering frequency f = 80...200 Hz. Thereby, the overlap is at least 50% (indentation distance a = 0.5 mm when indentation diameter d_i = 1 mm) in feed and path distance direction. By means of the following Equation (4.4), lower feed rates v_r = 40…100 mm·s^{-1} are calculated compared to those in Figure 4.2.

$$v_r = a \cdot f \quad \textbf{(4.4)}$$

It is therefore expected that a sufficiently large area with constant feed rates in a two-dimensional MHP processing will be set for the subsequent surface layer analyses. The adjusted robot feed rate is achieved over a sufficiently large distance. On this distance, the energy input into the workpiece caused by the feed rate, the hammering frequency and the plunger stroke is known. Thus, an evaluation of the surface layer properties takes place in this area due to the comparability of differently hammered conditions. In the following, the influence of the robot oscillations on the MHP processing result is examined additionally. Since the adjusted plunger stroke significantly determines the energy input into the workpiece, it is necessary to quantify a stroke change by robot oscillation in axial hammering direction.

4.1.2 Examination of robot oscillations

Untersuchung der Roboterschwingungen

During high-frequency MHP processing, oscillations occur along the impact axis. Due to the lower stiffness of industrial robots compared to machining centers, these influences must be quantified. For this purpose, MHP tests with hammering frequencies f = {120; 180} Hz were carried out along a measuring distance of 40 mm. The electrical power stage P_{el} = 70%, the stroke h = 0.4 mm and the indentation distance a = 0.05 mm were kept constant. These MHP parameters were used to set high hammering forces and the largest possible oscillations.

With higher material strengths, a greater effect on the oscillation behavior is expected due to the associated increased deformation resistance, thus determining the significance of the effects. Therefore, 100Cr6 bearing steel (DIN 1.3505, AISI 52100) was used as workpiece material because of the higher yield strength compared to AISI 304 stainless steel. Thereby, greater oscillations of the robot were expected. The dimensions of the specimens (length x width x height) were 100 x 40 x 30 mm. The chemical composition determined by means of spark spectroscopy meets the requirements of DIN EN ISO 683-17. The values of the element distribution (weight-%, wt-%) averaged over three measurements with the corresponding standard deviations σ (wt-%) are given in Table 4.1.

Table 4.1: Chemical composition of the used 100Cr6 bearing steel
Chemische Zusammensetzung des eingesetzten 100Cr6 Lagerstahls

Elements	C	Si	Mn	P	S	Cr	Mo	Cu
Weight-%	0.955	0.251	0.391	0.012	0.005	1.510	0.015	0.141
σ (wt-%)	0.019	0.001	0.003	<0.001	0.008	<0.001	0.003	<0.001

The hammered specimen, the plunger and the hammer tool oscillation in axial impact (z-) direction were investigated using an ARAMIS system from GOM GmbH, see Figure 4.3a. Based on triangulation, 3D displacements were optically recorded with 2,000 pictures per second in real time for point-based measurements [PICK13]. The programmed robot guaranteed a constant distance between the housing of the hammer tool and the workpiece. An analysis of the displacement of the hammer tool therefore allowed conclusions to be drawn about the robot oscillations. Using the GOM INSPECT software, the measured displacements were analyzed, see Figure 4.3b. By analyzing the three oscillating components, relative and absolute displacements were detected.

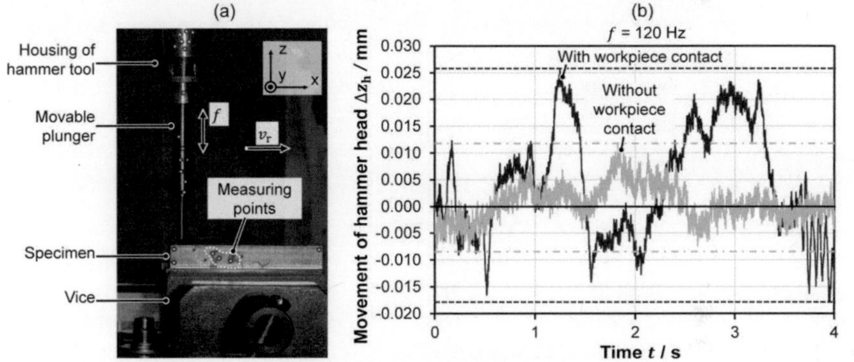

Test series: Robot oscillation **Workpiece material**: AISI 52100 **MHP-parameters**: d = 10 mm P_{el} = 70% h = 0.4 mm

Figure 4.3: Test setup with point markings using the GOM-ARAMIS system (a) and analysis of the hammer head movement for f = 120 Hz (b)
Versuchsaufbau mit Punktmarkierungen unter Verwendung des GOM-ARAMIS-Systems (a) und Analyse der Hammerkopfbewegung für f = 120 Hz (b)

Figure 4.3b shows the axial movement of the hammer tool at f = 120 Hz along one hammered path. The axial movement of the hammer tool is increased when MHP is performed with workpiece contact compared to the MHP without workpiece contact. The observed oscillations do not follow a regular pattern. In the parameter range considered, the maximum movement of the hammer tool was $\Delta z_h \approx 40$ µm. With activated MHP system, but without workpiece contact, a maximum movement $\Delta z_h < 20$ µm was detected. The maximum summed up oscillation of the robot was therefore 10% of the adjusted stroke. A comparison with KRALL's results on a KR30-3 robot from KUKA ($\Delta z_h \approx 250$ µm) makes it clear that the influences of different robots must be individually evaluated as an MHP carrier system. In addition to the robot vibrations, the oscillating plunger can also be elastically compressed in the axial impact axis. Therefore, finite element (FE) analyses were performed in SOLIDWORKS 2019, see Figure 4.4a.

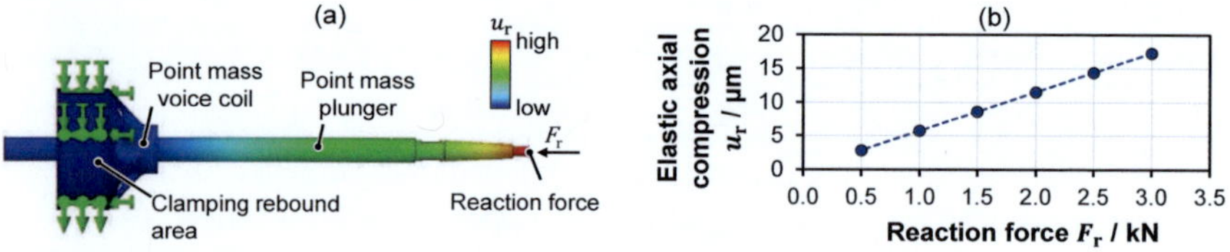

Figure 4.4: FE model (a) and related results (b) of elastic compression analysis of oscillating MHP mass using SOLIDWORKS software
FE-Modell (a) und zugehörige Ergebnisse (b) der Analyse der elastischen Stauchung der schwingenden MOH-Masse mittels SOLIDWORKS

The oscillating mass was modelled as 42CrMo4 (DIN 1.7225, AISI 4140) quenched and tempered steel. The YOUNG'S modulus E = 210 GPa, POISSON'S ratio ν = 0.3, density $\rho = 7.72 \cdot 10^{-6}$ kg/mm³ and yield strength $R_{p0.2}$ = 800 MPa were used [DEW15]. The reaction forces were varied between F_{reac} = 0.5...3.0 kN. This force range was identified in preliminary tests as relevant for electrodynamic MHP with the ACCURAPULS system type 2002. The simulation procedure was based on the work of HABERSOHN [HABE15]. No plastic deformation was observed. Maximum elastic compressions $u_{r,max} \approx$ 5...20 µm were determined, see Figure 4.4b. The resulting axial stiffness of the vibrating mass c_{ax} = 167 N/µm corresponds to HABERSOHN's results [HABE15].

The influences on the surface topography resulting from the robot oscillation and the elastic compression of the oscillating mass were investigated in the following. For this purpose, 20x20 mm fields on 100Cr6 workpieces were hammered in a meandering pattern. By repeated wavelet transformations, the output profile was decomposed to make the waviness visible, see Figure 4.5.

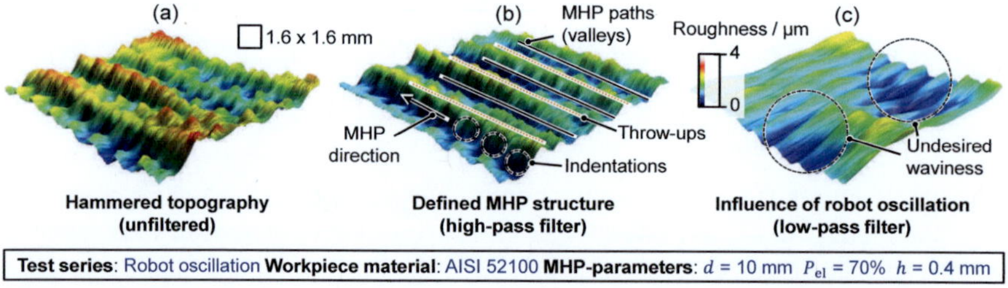

Figure 4.5: Influence of robot oscillation on surface topography during MHP: unfiltered topography (a), defined MHP structure (b) and undesired waviness (c)
Einfluss der Roboterschwingung auf die Oberflächentopografie beim MOH: Ungefilterte Topografie, definierte MOH-Struktur (b) und ungewünschte Welligkeit (c)

The slight oscillation of the carrier system (robot axis, hammer tool) with the MHP tool can lead to undesired surface effects. In Figure 4.5b, the defined individual indentations of the MHP head are visible. The individual hammer paths define the specific surface structure. By keeping a large distance between the individual hammer paths, the material throw-ups can be specifically reduced. In Figure 4.5c, however, the waviness caused by the robot oscillation is undesired and prevents a defined surface structuring. Such a pattern was found at several locations on the hammered specimen. After the robot oscillates in positive axial hammering direction, it over swings in negative direction. Nevertheless, the size range of the unintentionally induced waviness due to the robot oscillation is very small. The surface topography determined after MHP tests in Chapter 6.1.3 support this statement. In addition to the influence of axial robot oscillation on MHP processing, deviations in the processing plane can also occur. These deviations in MHP feed and path change direction will be presented in the following.

4.1.3 Influence of robot dynamics on MHP processing

Einfluss der Roboterdynamik auf die MOH-Bearbeitung

Using a micro-structured plunger with a hammer head diameter d = 10 mm, the machining deviations due to oscillations of the carrier system parallel (x) and perpendicular to the hammering direction (y) were investigated. The structured area on the hammer head had a diameter d_{m} = 100 µm and a depth z_{m} = 2 µm, see Figure 4.6a.

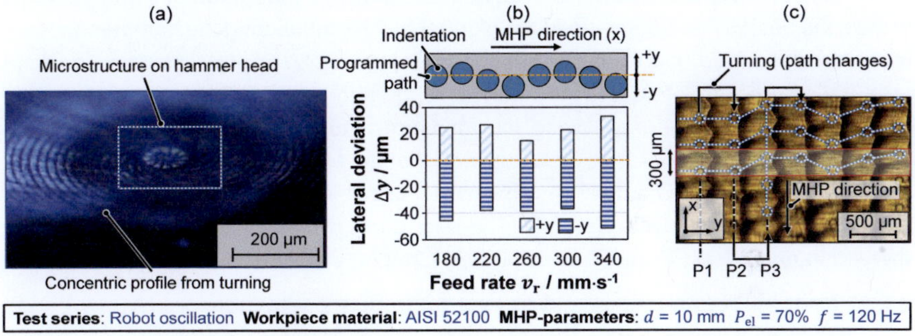

Figure 4.6: Microstructured hammer head surface (a), deviation of indentations along feed axis (MHP direction) (b) and process-related deviation of indentations of adjacent MHP paths (c)

Mikrostrukturierte Hammerkopfoberfläche (a), Abweichung der Eindrücke entlang der Vorschubachse (MOH-Richtung) (b) und prozessbedingte Abweichung von Eindrücken nebeneinanderliegender MOH-Bahnen (c)

Using a hammering frequency f = 120 Hz, it was found that the indentations in the feed direction (x) were equidistant and in line. Assuming that the indentation diameters are about d_{i} = 1 mm, the maximum lateral deviation Δy_{max} = 50 µm can be neglected. A high positional repeatability of the robot is therefore given. However, adjacent paths (y-direction) show a partly significant displacement, which is due to the turning points during path changes with meandering machining. An exact positioning of the individual indentations of adjacent paths is not possible due to the process characteristics [LECH14]. In some cases, larger deviations can occur $\Delta y_{\mathrm{max}} \approx$ 300 µm. These deviations are still within the range of a single indentation diameter and therefore negligible.

The presented results show the general suitability of the used industrial robot as MHP carrier system. The influence in axial hammering direction and thus on the energy input is small. The deviations in the machining plane do not lead to a significant influence on the energy density introduced in a two-dimensional MHP machining process.

In the following, the results of the metrological extension of the electrodynamic MHP system are presented. The data recorded in this way allow conclusions to be drawn about the process kinetics and the energy introduced into the processed workpiece surface layer. The knowledge of the MHP energy introduced is essential for the determination of the mechanisms of action in the surface layer of hammered materials.

4.2 Metrological extension of the electrodynamic MHP system

Messtechnische Erweiterung des elektrodynamischen Hämmersystems

In this subsection, the results of the metrological extension of the used electrodynamic MHP system are presented. In a first step, a requirement analysis for the inline sensors was done. Since electrical energy is converted into kinetic energy in the electrodynamic MHP system, both electrical (voltage, current) and mechanical (stroke, hammer force) variables were recorded. The implementation and calibration of the sensors as well as the design developments of the plunger tool for measuring the hammering force were also described. The recorded sensor data were analyzed and discussed in Chapter 4.3. Of particular interest is the analysis of the stroke, since it has a decisive influence on the contact energy. Using the measured values and machine learning approaches, a numerical MATLAB/SIMULINK model of the electrodynamic MHP was developed and validated in Chapter 4.4. This model allows a reliable prediction of the contact energy for electrodynamic MHP. The recorded process kinetics were also implemented in numerical FE analyses of MHP in Chapter 6.2.

4.2.1 Requirement analysis and sensor technology selection

Anforderungsanalyse und Auswahl der Sensortechnologie

According to the design methodology of PAHL and BEITZ [BEND20], which is based on VDI guideline 2221 [VDI93], the three main phases of specification, design and construction ensure a systematic and comprehensive tool development. Prior to the implementation of force and displacement sensor in an electrodynamic MHP system type 2002 from ACCURAPULS GMBH, a requirement analysis was carried out using a specification sheet. All identified requirements were divided into necessary (N) and desirable (D) requirements. In preliminary studies on electrodynamic MHP on cuboid AISI 304 workpieces with a KISTLER 9257B force measurement platform, a maximum axial hammering force F_h = 3.0 kN was determined. This finding was taken into account in the requirements analysis for the inline force sensor. Table 4.2 shows an extract from the specification sheet.

Table 4.2: Extract of the requirement analysis for sensors in electrodynamic MHP
Auszug der Sensor-Anforderungsanalyse beim elektrodynamischen MOH

Feature	Necessary (N) / Desirable (D)	Requirement
Force	N	Continuous inline measurement
	N	Close to forming zone
	N	No influence on process kinematics
	N	Axial force measurement (F_{h} = -3...3 kN)
	N	High sensor stiffness and eigenfrequency
	D	Integrated impedance converter
	D	Low force bypass
Stroke	N	No influence by magnetic field
	N	High resolution (Δh_{S} < 1 μm)
	N	Measuring range (h_{S} = 0...2 mm)
	N	No influence on process kinematics

With the specification sheet, a fiber-optic reflection-compensated distance sensor RC20 from PHILTEC with a measuring range $h_{\mathrm{RC20}} \leq 1.65$ mm and a resolution Δh_{RC20} = 0.6 μm was selected. A precise measurement in small stroke ranges with a high resolution is essential for the electrodynamic MHP. In most industrial applications, a stroke h < 1 mm is set. Due to the fiber optic measuring principle, an influence on the oscillating mass was also avoided. An IEPE piezo force sensor 9712B5000 from KISTLER with a mass m_{B5000} = 19 g, integrated impedance converter electronics and a measuring range ΔF_{ax} = -2.22...22.2 kN in axial hammering direction was chosen for force measurement. Thus, the measuring chain was reduced. Fewer cables automatically mean less interference in signal acquisition. The measuring range is also suitable for short-term load peaks during MHP processing. After the sensor technology was selected based on the requirement analysis, the hammer system was extended accordingly by integrating the sensor technology.

4.2.2 Instrumentation of MHP tool

Aufrüstung des MOH-Werkzeugs

After selecting the sensor technology, both sensors were implemented in the hammer tool. Figure 4.7a shows the schematic design of the hammer tool and the mounting position of the distance sensor. The sensor works according to the principle of reflected intensity. Since the oscillating mass in the hammer tool can be regarded as a coherent unit, the stroke exerted on the workpiece is determined by means of a back-sided sensor placement, see Figure 4.7b. By embedding a polished stainless steel plate in epoxy resin above the oscillating mass, optimal reflection conditions were created for the distance sensor, see Figure 4.7c.

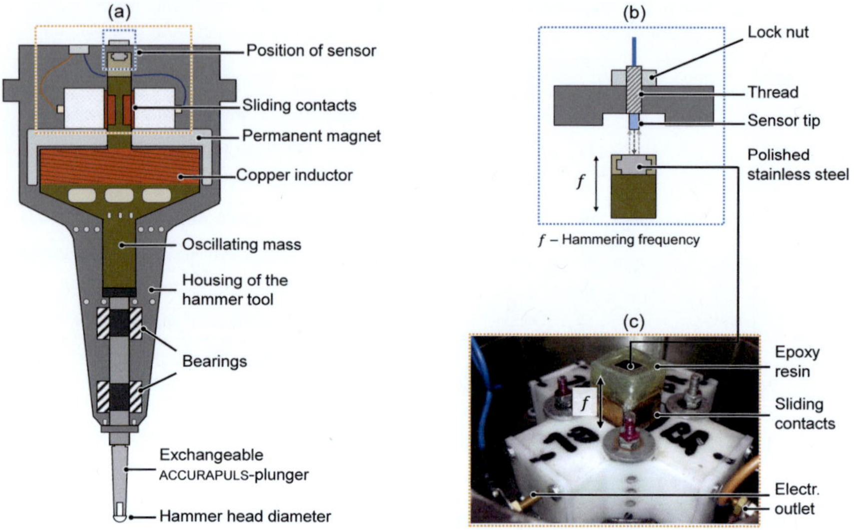

Figure 4.7: Schematic representation of the electrodynamic hammer tool (a), mounting position of the distance sensor (b) and embedded stainless steel plate (c)
Schematische Darstellung des elektrodynamischen Hämmerwerkzeugs (a), Befestigungsposition des Abstandssensors (b) und eingebettete Edelstahlplatte (c)

For the best possible reflection conditions, a thin stainless steel plate polished to Sa = 0.3 µm was cast into an epoxy resin matrix at the upper end of the oscillating body, see Figure 4.7c. The additional mass m_{er} = 4 g, compared to the mass of the oscillating body m_{ob} = 363 g together with the plunger mass m_{p12} = 70.1 g (hammer head diameter d = 12 mm) or m_{p6} = 77.2 g (d = 6 mm), has a negligible influence on the MHP process kinematics.

The sensor was calibrated by means of programmed vertical movement of an ABB IRB6660-205/1.9 robot using a compensation curve and a high-speed camera PHANTOM v7.3. The procedure and calibration results are shown in Figure A.1 in the appendix. After calibration, the stroke h was determined with an accuracy $\Delta h = \pm$ 1 µm in the operational range h = 0.25…1.7 mm. The software LABVIEW and DIADEM from NATIONAL INSTRUMENTS were used for data acquisition, visualization and evaluation. The impact velocities derived from the stroke data multiplied by the oscillating mass were used to calculate the kinetic energy E_{kin} which was defined as contact energy E_{c}.

Figure 4.8a and b show the mounting position and the dismantled ACCURAPULS plunger. These plungers consist of a solid carbide ball pressed into a hardened steel shaft. The developed force-measuring plunger is shown in Figure 4.8c. With a total mass m_{FP} = 73.5 g, there is a deviation from the ACCURAPULS plunger Δm < 5%.

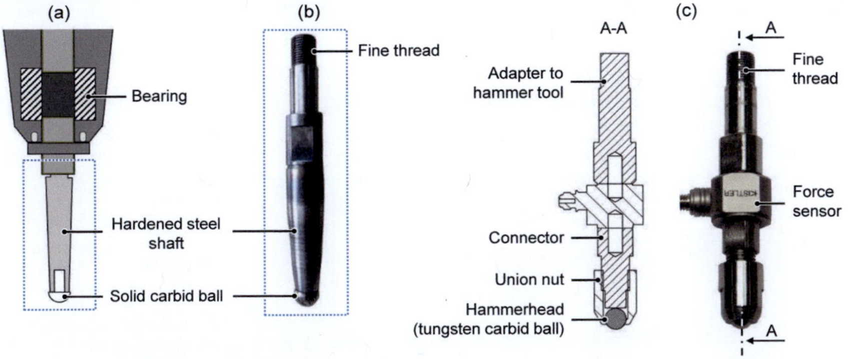

Figure 4.8: Detailed view (a) and dismantled ACCURAPULS plunger (b); developed force-measuring plunger (c)
Detailansicht (a) und zerlegter ACCURAPULS-Stößel (b); entwickelter Kraftmessstößel (c)

To avoid plastic deformation during MHP, carbide balls made of CTS20D from CERATIZIT with a hardness of 1,600 HV30 and a surface roughness Sa < 0.2 µm were used. The smooth ball surface prevents any influence on the workpiece roughness with MHP. The adapter to the hammer tool, the connector and the union nut were made of quenched and tempered steel 42CrMo4 (DIN 1.7225, AISI 4140).

The force-measuring plunger was designed in a modular way so that an almost constant plunger length and weight is maintained for different ball diameters. Repeated set-up of the MHP process is facilitated by the modular plunger design. Repetitive hammering with the force-measuring plunger allows the carbide ball to indent a recess for an optimum fit on the connector. The proximity of the force sensor to the forming point allows an almost undistorted force measurement. The force-measuring plunger was calibrated using a dynamometer 9257B from KISTLER. By means of the developed force measuring plunger, the hammering forces can be determined directly in the process precisely and with little metrological effort, regardless of the workpiece geometry.

After installation of the (further developed) sensor technology in the hammer tool, the effects of the sampling rate on the measurement result were first investigated. The electrodynamic MHP is characterized by very short contact times t_c = 0.5…1.5 ms depending on the stroke [KRAL19]. Therefore, sufficiently high sampling rates must be selected for valid and precise measurement results of the inline sensors.

4.2.3 Influence of sampling rate on process data

Einfluss der Abtastrate auf die Prozessdaten

Due to the dynamic process characteristics and the small maximum plunger stroke $h_{\max}$ < 2 mm, the sampling rate required for sufficient resolution of the hammering force and plunger stroke was investigated. The MHP was performed with a hammering frequency f = 90 Hz and a plunger stroke h = 0.3 mm on a pre-strengthened point on an X5CrNi18-10 (DIN 1.4301, AISI 304) workpiece. A further plastic deformation and thus falsification of the hammering force could be excluded according to TRAUTH [TRAU13]. Figure 4.9a shows the influence of different sampling rates on the axial hammering forces at the workpiece and the internal stop in the hammer tool. The oscillating mass in the hammer tool alternates between the workpiece and an inner stop.

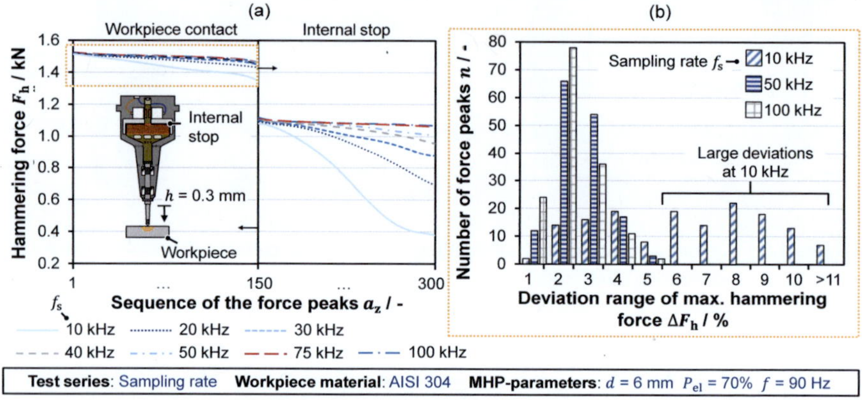

Figure 4.9: Influence of sampling rate on hammering force (a) and ranges of deviation (b)
Einfluss der Abtastrate auf die Hämmerkraft (a) und Abweichungsbereiche (b)

Taking into account the WKS sampling theorem, the range of the sampling frequency f_s = 10...100 kHz was investigated [WHIT28]. After the hammering system had settled in after a few minutes, 150 hammering cycles were analyzed. A force peak during workpiece contact and a force peak at the internal stop characterize each hammering cycle. The force peaks were sorted and displayed from highest to lowest. The sloping curve in workpiece contact and the internal stop at a sampling frequency f_s = 10 kHz is evident, see Figure 4.9a. At the other sampling frequencies, only a slightly decreasing curve is visible. This indicates a lower scattering of the force maxima.

Figure 4.9b shows the number of the force peaks and their deviation from the maximum measured hammering force. It can be seen that at a sampling frequency f_s = 10 kHz a large scattering with great deviations of the force values occurs. By means of a sampling rate f_s = 50 kHz all force peaks occurring within a maximum deviation $\Delta F_h \leq 5\%$ can be determined. An increase of the sampling rate to f_s = 100 kHz leads to a slightly smaller scattering widths, but to a significantly increased amount of data. It is therefore assumed that a sampling frequency f_s = 2 kHz selected by TRAUTH [TRAU16A] was chosen to small and leads to a distortion of the hammering forces in electrodynamic MHP. Some force peaks were therefore not recorded. A sampling rate f_s = 50 kHz could also be determined for the valid recording of the plunger stroke. After implementing and calibrating the inline sensors and determining the required sampling rate, the effects of selected MHP process parameters on the robot-guided MHP system behavior were investigated.

4.3 Effects of process parameters on MHP system behavior

Einfluss der Prozessparameter auf das MOH-Systemverhalten

In this chapter, the data recorded by means of the implemented force and distance sensors from Chapter 4.2 were evaluated. In addition to the maximum free stroke and stroke curves for different MHP parameter setups, the hammering force and electrical input values were determined. The maximum free stroke as a function of the hammering frequency defines the adjustable MHP parameter range in Chapter 6.1 and 7.1. The electrical input variables were used for a MATLAB/SIMULINK process model of the electrodynamic MHP in Chapter 4.4. The contact energies served as input to a finite element (FE) process model of the MHP in Chapter 6.2.

4.3.1 Test setup and experimental procedure

Versuchsaufbau und experimentelle Vorgehensweise

The process kinematics, the electrical input variables and the hammering forces of the electrodynamic MHP system type 2002 from ACCURAPULS were determined by means of MHP tests. The tests were carried out on a pre-strengthened point on an AISI 304 specimen with a robot ABB IRB6660-205/1.9. The cuboid specimen with the dimensions 100 x 40 x 30 mm (L x W x H) was clamped centrally on a machine table using a vice. The MHP was performed in the center of the specimen. An influence of the measurement results by a movement of the specimen could thus be avoided. All measurements were performed after 5 minutes in steady state.

The hammering frequency f = 20...400 Hz in 20 Hz steps, the electrical power stage P_{el} = 40...80% in 10% steps, the stroke h = 0.1...1.0 mm in 0.1 mm steps and the hammer head diameter d = {6; 12} mm were varied. The implemented and calibrated force and distance sensors were used for measurement of the hammering force and the plunger stroke. The force measurement chain also included a multi-channel charge amplifier 5070 from KISTLER. Using a current input module 9247 and a voltage input module 9225 from NATIONAL INSTRUMENTS, the current and the voltage were measured. The voltage and current signals were recorded between the power electronics in the control cabinet and the hammer tool. All variables were acquired and visualized using the LABVIEW software with a sampling rate f_s = 50 kHz. The evaluation of the data was done using the DIADEM software from NATIONAL INSTRUMENTS.

4.3.2 Free stroke analysis

Freihubanalyse

As already mentioned, the knowledge of the maximum free stroke in dependence of the hammering frequency is important for the interpretation of a meaningful process window in Chapter 6.1. The free stroke is defined as stroke without workpiece contact. The stroke graphs for hammering frequencies f = 20...400 Hz were determined by means of the implemented distance sensor and a high-speed camera PHANTOM v7.3. Since the distance sensor only measures in the stroke range h = 0.25...1.70 mm, the stroke was determined for the adjacent areas using a high-speed camera. The curves for the two hammer head diameter d = {6; 12} mm are shown in Figure 4.10a.

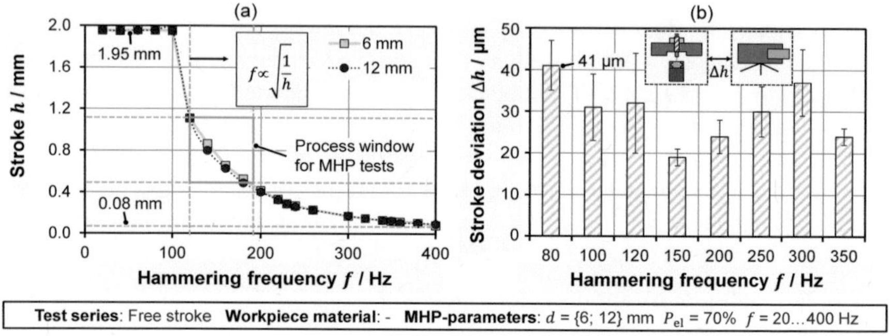

Figure 4.10: Free stroke measurement (a) and deviation between distance sensor and high-speed camera (b)
Freihubmessung (a) und Abweichung zwischen Abstandssensor und Hochgeschwindigkeitskamera (b)

There are no significant differences between the two graphs in Figure 4.10a. The difference in weight of 7 g between the two plungers has no effect on the oscillating mass. The electrical power stage was set to P_{el} = 70%. Up to a hammering frequency f = 100 Hz, the maximum free plunger stroke h_{max} = 1.955 mm is achieved. Up to a hammering frequency f = 400 Hz, the free plunger stroke drops to a minimum value h_{min} = 0.080 mm. Due to the mode of action, the maximum plunger stroke can no longer be achieved at higher hammering frequencies. The cause therefore is that the number of stroke cycles per second increases. The curve between f = 120...400 Hz can be described with the Equation (4.5). These finding confirms the results of TRAUTH [TRAU16A] on a ACCURAPULS electrodynamic hammer system of the previous generation with less electrical power.

$$f \propto \sqrt{\frac{1}{h}} \tag{4.5}$$

To ensure the compatibility of the high-speed camera and sensor data, the measurement deviations between them were determined, see Figure 4.10b. The data recorded by the distance sensor in the examined frequency range were $\Delta h \approx$ 20...40 µm above the values determined by the high-speed camera. Due to difficulties in tracking individual pixels in the high-speed recordings, these deviations occurred. The highest deviations occurred for a hammering frequency f = 80 Hz. Since the deviations were only small, these camera data was used to create the fully free stroke profile.

A compromise between processing time (depending on hammering frequency) and available plunger stroke for the MHP tests in Chapter 6.1 was derived from the results. Using a hammering frequency f = 120...180 Hz (orange area in Figure 4.10a), the system behavior can thus be investigated over a wide stroke and contact energy range.

4.3.3 Determination of current, voltage and electrical power

Ermittlung des Stroms, der Spannung und der elektrischen Leistung

Since in electrodynamic MHP, the electrical input variables current and voltage influence the mechanical domain and thus the contact conditions with the workpiece, these were investigated. Due to only small differences in plunger weight, similar values were determined for the hammer head diameters d = {6; 12} mm. Figure 4.11a shows the input voltage U for different power stages P_{el} = 40...80%. The hammering frequencies f = 20...400 Hz were varied and the plunger stroke h = 0.1 mm was constant. Because the stroke was chosen so small, the entire hammering frequency range could be examined and extensive characterization could be carried out.

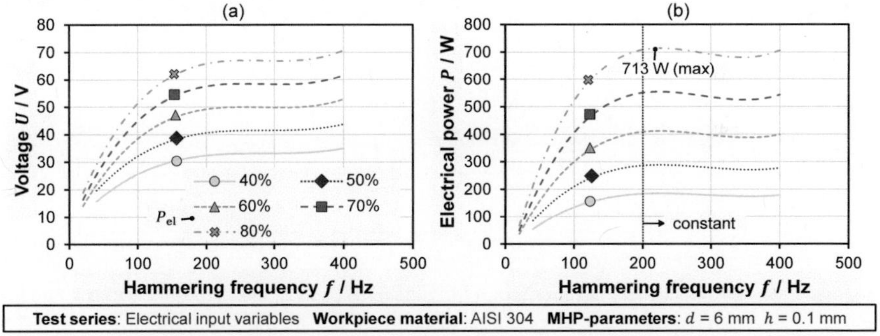

Figure 4.11: Measured voltage for different power stages and hammering frequencies (a) and electrical power calculated from current and voltage (b)
Gemessene Spannung für verschiedene Leistungsstufen und Hämmerfrequenzen (a) und aus Strom und Spannung berechnete elektrische Leistung (b)

In preliminary tests, it could be determined that an increase in stroke has no significant effect on the current and voltage. The input voltage increases up to a hammering frequency f = 200 Hz for all power stages and then remains nearly constant up to a frequency f = 400 Hz. A comparison with the results of TRAUTH [TRAU16] shows that the magnitude of the electrical input of different system generations differ considerably. The electrical power P was calculated using the input current (appendix, Figure A.2) and is shown in Figure 4.11b.

As the electrical MHP power level increases, the measured current level also increases. The maximum current values were measured at hammering frequencies f = 120...140 Hz. The underlying current values were between I = 2.50...10.90 A. The electrical power shows a similar trend with a maximum at a hammering frequency around f = 200 Hz. The highest electrical power P_{max} = 713 W is determined for the MHP power stage P_{el} = 80%. Thus, the electrical power is more than thrice as high as the measured values in [TRAU16A]. Due to the higher electrical power compared to TRAUTH [TRAU16], higher hammering forces were expected. These are quantified in the following.

4.3.4 Determination of hammering forces

Bestimmung der Hämmerkräfte

The hammering forces as a function of the hammering frequencies are shown in Figure 4.12a and as a function of the plunger stroke are shown in Figure 4.12b. The hammering forces were measured by means of the developed force-measuring plunger and verified punctually with a dynamometer.

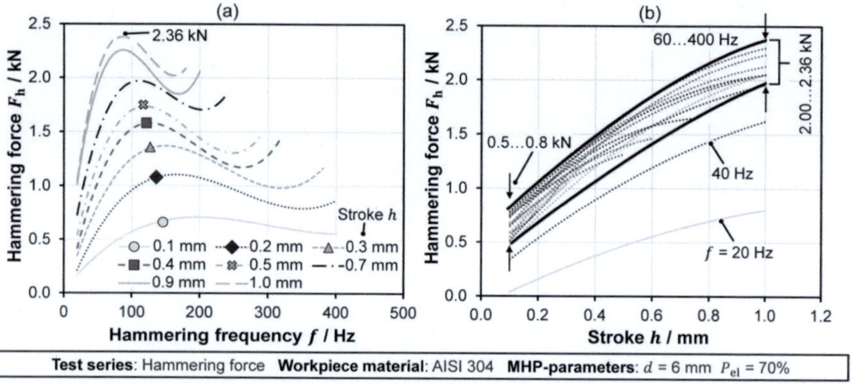

Figure 4.12: Evaluation of the hammering force as a function of the hammering frequency (a) and development of the hammering force with increasing stroke (b)
Bestimmung der Hämmerkraft in Abhängigkeit von der Hämmerfrequenz (a) und Verlauf der Hämmerkraft mit steigendem Hub (b)

The hammering force increase with increasing plunger stroke h = 0.1…1.0 mm. The increase of the hammering force is not linear. As the stroke increases, the maximum hammering force shifts towards lower hammering frequencies. The greatest hammering force $F_{h,max}$ = 2.36 kN was achieved with a plunger stroke h = 1.0 mm at a hammering frequency f = 90 Hz. Figure 4.11b shows that the greatest hammering forces do not occur at the highest measured electrical power. One possible explanation is the increasing share of loss terms such as friction and heat in the hammer tool with increasing hammering frequency.

Figure 4.12b shows that significantly lower maximum hammering forces are obtained with small hammering frequencies f = {20;40} Hz. The range of hammering frequencies f = 60…400 Hz has been framed for reasons of clarity. In this range, the hammering forces for a small stroke h = 0.1 mm are between F_h = 0.5…0.8 kN. For a large stroke h = 1.0 mm, the forces are $F_{h,max}$ = 2.00…2.36 kN. The higher electrical input power of the electrodynamic MHP system type 2002 results in significantly higher hammering forces compared to older generation systems ($F_{h,max,old}$ = 1 kN) [TRAU16A]. Furthermore, the free stroke set by the manufacturer has a significant influence on the acceleration path of the MHP plunger and thus on the kinetic energy. The measurements have clearly shown the deviations between systems of different generations. However, the basic MHP system behavior is very similar.

4.3.5 Determination of plunger stroke and contact energy

Bestimmung des Stößelhubes und der Kontaktenergie

The stroke and thus the acceleration distance of the oscillating mass has a significant influence on the contact energies in electrodynamic MHP. As a result, the surface layer properties of the hammered workpiece can be adjusted over a wide range. Depending on the operating principle and process settings, however, the stroke characteristics can differ. In Figure 4.13a, the influence of different MHP electrical power stages P_{el} = 50...80% on the course of two impacts is shown. A hammer head diameter d = 6 mm and a stroke h = 0.5 mm were chosen.

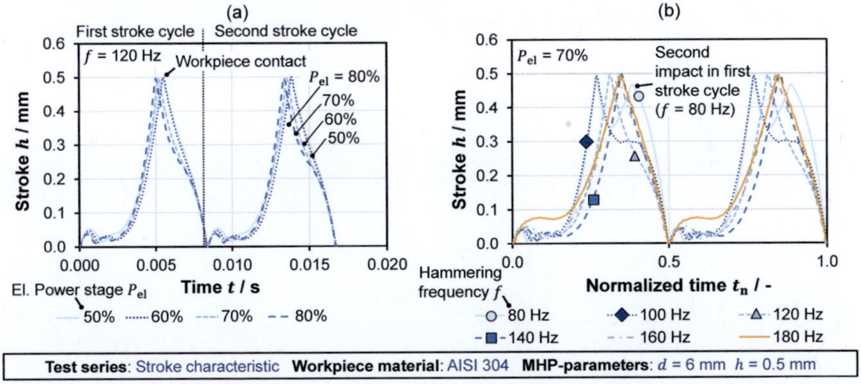

Figure 4.13: MHP kinematics for a hammering frequency f = 120 Hz at different electric power stages P_{el} (a) and for a stroke h = 0.5 mm for different frequencies f (b)
MHP-Kinematik für eine Hämmerfrequenz f = 120 Hz bei verschiedenen Leistungsstufen P_{el} (a) und für einen Hub h = 0,5 mm bei verschiedenen Frequenzen f (b)

The graphs in Figure 4.13a are almost on top of each other and only minimal differences are visible. A lower electrical power level ensures a more symmetrical shape of the impact. However, this only refers to the rebound path of the plunger away from the workpiece. By reducing the electrical power level, contact with the workpiece only occurs at a later stage Δt = 0.0005 s. This corresponds to a time delay of 6% in relation to a stroke cycle. The same observations were also made for other strokes h and hammering frequencies f. The influence of the electrical power stage in the considered parameter range on the impact characteristic is therefore small.

Figure 4.13b shows selected stroke characteristics with a power stage P_{el} = 70%. The set stroke is maintained very precisely. Before the actual stroke cycle is executed, the oscillating mass undergoes a first slight oscillation movement, with the exception of the hammer frequency f = 180 Hz. At a low frequency f = 80 Hz, a second impact occurs after the first contact of the hammer head with the workpiece surface during one stroke cycle. With increasing frequency this effect decreases further and a symmetrical stroke before and after the workpiece contact occurs. As the frequency increases, the hammer head comes into contact with the workpiece with a delay. The acceleration of the hammer head between the frequencies f = 80...140 Hz is similar (comparable gradient of the graphs) and decreases for the frequencies f = 160...180 Hz.

Since obviously with some MHP parameter settings multiple impacts can occur within one stroke cycle, this phenomenon was considered more closely in the following. For an electrical power stage P_{el} = 70%, a hammer head diameter d = 6 mm and a hammering frequency f = 120 Hz, the influence of different strokes h on the impact characteristic has been investigated, see Figure 4.14. The force measurements were included to clearly determine multiple impacts. The normalized force and stroke graphs of an impact cycle were plotted over the normalized time.

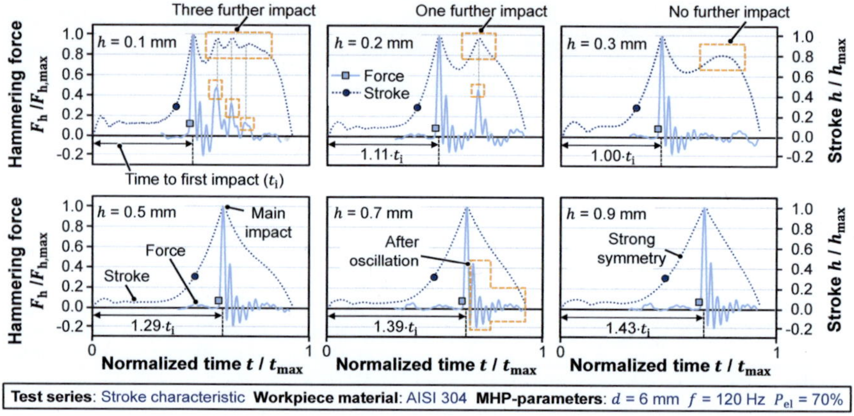

Figure 4.14: Analysis of impact curves for one stroke cycle and different strokes
Analyse von Einschlagverläufen für einzelnen Schlagzyklus und verschiedene Hübe

At a small stroke h = 0.1 mm, three additional impacts were detected. At a stroke h = 0.2 mm one additional impact occurs. As a result, it is not possible to set an optimum impact at small strokes. Based on the force data for a stroke h = 0.3 mm, a supposed second impact could be excluded. From a stroke of $h \geq$ 0.5 mm, the impact profile approaches a symmetrical course. All curves show a similar characteristic area immediately after the main impact, which indicates after oscillations in the force measurement. Furthermore, a delay of the impact time with increasing stroke was detected. The greater free path length of the plunger with the same oscillating mass and available system power can explain this. The result is also a shorter contact time of the plunger with the workpiece and thus better conditions for optimized energy conversion. All determined stroke graphs show a clearly deviating behavior from an idealized cosine movement. Depending on the MHP process parameters, very different stroke characteristics occur. This in turn results in different contact conditions and energies at the workpiece surface layer and thus changed workpiece properties after MHP.

After analyzing the impact characteristics, DIADEM software was used to determine the impact velocities by deriving the stroke profiles. Within these curves, three sections with 10 stroke cycles each were selected. From the total thirty stroke cycles, the maximum impact velocity was determined each per stroke and the mean value was then calculated. The graphs were smoothed with a third-degree polynomial function. A coefficient of determination $R^2 > 95$ was obtained for all curves. Figure 4.15a shows the impact velocities for the hammer head diameter d = 12 mm at the electrical power stage P_{el} = 70% for various stroke-frequency combinations.

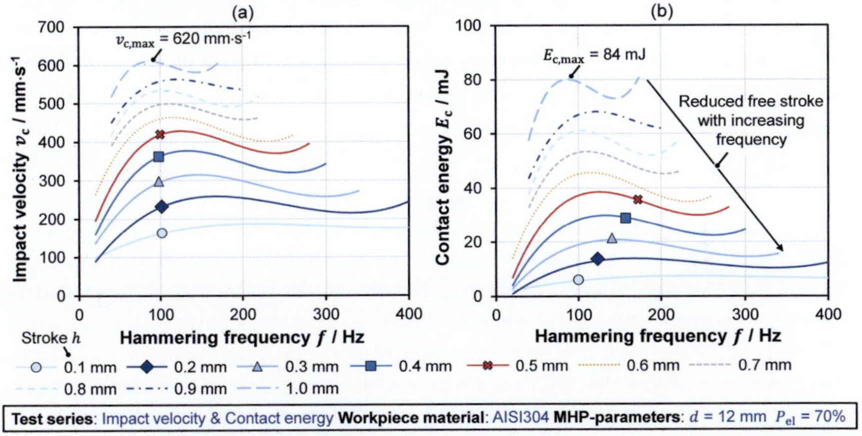

Figure 4.15: Evaluation of impact velocities (a) and calculated contact energies (b) as function of hammering frequency and stroke
Auswertung von Aufprallgeschwindigkeiten (a) und berechneten Kontaktenergien (b) als Funktion der Hämmerfrequenz und des Hubes

When the stroke h = 0.1...1.0 mm increases, the impact velocity increases along the hammering frequency band. A maximum impact velocity $v_{\mathrm{c,max}}$ = 620 mm·s^{-1} is achieved with a stroke h = 1.0 mm and a frequency f = 180 Hz. As the stroke increases, the maximum impact velocity shifts towards lower hammering frequencies. These findings confirm the results of TRAUTH [TRAU16A]. The range of impact velocities relevant for the parameter setups selected for the MHP tests, which will be explained in further chapters, lies between v_{c} = 300...420 mm·s^{-1}. The impact velocities determined by means of the fiber-optic distance sensor are also in the same order of magnitude as the impact velocities determined by KRALL by means of an eddy current sensor on a younger generation ACCURAPULS hammer system [KRAL19]. By knowing the weight of the oscillating mass and the masses of the plunger, the contact energy E_{c} was calculated using the following Equation (4.6):

$$E_{\mathrm{kin}} = E_{\mathrm{c}} = \frac{1}{2} \cdot (m_{\mathrm{ob}} + m_{\mathrm{p}}) \cdot {v_{\mathrm{c}}}^2 \quad \textbf{(4.6)}$$

with m_{ob} - mass of the oscillating body, m_{p} - mass of the plunger and v_{c} - impact velocity.

The calculated contact energies are shown in Figure 4.15b. Similar to the impact velocities, the maximum contact energy shifts to lower hammer frequencies as the stroke increases. The maximum contact energy in the investigated parameter range and electrical power level P_{el} = 70% is $E_{c,max}$ = 84 mJ. Due to the increasing stroke, the acceleration path of the plunger and thus the contact energy increase. The deviations between the contact energies at d = 6 mm and d = 12 mm are about ΔE_c = 5%. It must be mentioned that the contact energy determined here corresponds to the kinetic energy of the oscillating mass in the hammer tool. There is no division into an elastic and a plastic energy component. However, since the plastic energy component already predominates for low impact velocities and low masses, this approach is chosen as a first approximation.

The above-mentioned experiments were used to determine the impact velocities and calculate the contact energies over a wide MHP-parameter range. It was shown that electrodynamic hammer systems of various generations differ in their kinetics with similar system behavior. A general correlation between the MHP process variables (d, f, P_{el}) and the resulting impact velocity would enable a fast estimation of the impact velocity independently of the MHP system. With the additional knowledge of the cause-and-effect relationships between introduced contact energy and the resulting surface layer properties after MHP, an initial rapid process design would thus made possible in the future. For this reason, a correlation between the MHP process parameters and the impact velocity is derived in the following.

4.3.6 Deduction of an empirical-mathematical MHP model

Herleitung eines empirisch-mathematischen MOH-Modells

From the experimentally determined stroke and the calculated impact velocities, a correlation was derived which is generally valid over the analyzed MHP parameter range. This serves as a quick and approximate estimation of the impact velocity in electrodynamic MHP. For this purpose, selected function types were first implemented in the curve fitting toolbox in MATLAB calculation software. By means of the numerical optimization algorithm according to LEVENBERG [LEVE44] and MARQUARDT [MARQ63], the experimental impact velocity data were approximated. These algorithms were developed to solve non-linear fitting problems using the method of non-linear least squares. The function with the best fit to the experimental data was subjected to nonlinear optimization using the trust region reflective algorithm to approximate the numerical model parameters [GELE07]. The empirical-mathematical model for predicting the impact velocity v_c [mm·s^{-1}] is shown in Equation (4.7):

$$v_c = a_1 \cdot {P_{el}}^{a_2} \cdot h^{a_3} \cdot (a_4 + h^{a_5} \cdot \ln(f)) \qquad \textbf{(4.7)}$$

with a_{1-5} as numerical constants, P_{el} [%] the electrical power stage of the MHP system, h [mm] the plunger stroke and f [Hz] the hammering frequency. The numerical constants were determined as follows: a_1 = 11.9239, a_2 = 0.5087, a_3 = -0.09335, a_4 = 1.00 und a_5 = 1.0952.

The impact velocities for electrodynamic MHP were calculated with the empirical-mathematical Equation (4.7) and compared with the experimental values for the electrical power stage P_{el} = 70%, see Figure 4.16a. The results in Figure 4.16b are an extract from Figure 4.16a.

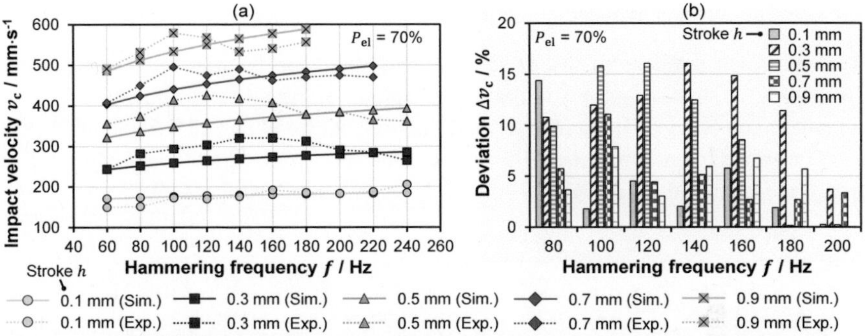

Figure 4.16: Model validation by means of comparison of calculated and experimentally determined impact velocities (a) and detailed analysis of deviations (b)
Modellvalidierung durch Vergleich der berechneten und experimentell bestimmten Aufprallgeschwindigkeiten (a) und Detailanalyse der Abweichungen (b)

The prediction accuracy for a low plunger stroke h = 0.1 mm is good. At hammering frequencies f = 100...220 Hz, the deviations are $\Delta v_c \leq 5\%$. This is the standard frequency range in the industry. For almost all hammering frequencies and a stroke h = 0.3 mm the impact velocity is significantly underestimated ($\Delta v_c > 10\%$) by means of the developed model. For a stroke h = 0.5 mm, a similar trend is observed with mostly Δv_c = 5...10%. In principle, a more precise prediction of the impact velocity can be made with the model for higher strokes. The prediction accuracy of the model is generally higher for low electrical power levels than for higher power levels. The model is able to predict impact velocities with a mean percentage deviation of 8% in the considered parameter space. Nevertheless, the impact velocities calculated using the model deviate in some cases significantly from the experimental values. The representation of such a complex, multiphysical system behavior is only possible to a limited extent with the present empirical-mathematical model. One way to reproduce this behavior more precisely is to use numerical methods. Therefore, based on the work of TRAUTH [TRAU16A] and HABERSOHN [HABE15], a process model of electrodynamic MHP was further developed using the software MATLAB/SIMULINK.

4.4 Numerical modelling of multiphysical MHP system behavior

Numerische Modellierung des multiphysikalischen MOH-Systemverhaltens

In the previous subsection, it was shown that the experimentally determined system behavior of electrodynamic MHP could only be described in a limited way by means of a simple general correlation. For this reason, it is reasonable to model the complex multiphysical system behavior numerically by means of MATLAB/SIMULINK. In the following, the electrical, mechanical and contact domains of electrodynamic MHP, which are decisive for the system behavior, will therefore be examined more closely.

4.4.1 Electrical operating system

Elektrisches Wirksystem

The operating principle of the electrodynamic MHP has already been explained in Chapter 2.3.2. In the center of the electrical domain is the model of a plunger coil. The induction coil (inductor), the plunger and the hammer head describe a coherent oscillating mass in the hammer tool. This mass immerses into the permanent magnet after applying an alternating sinusoidal voltage $U_0(t)$ to the inductor, see Figure 4.17a.

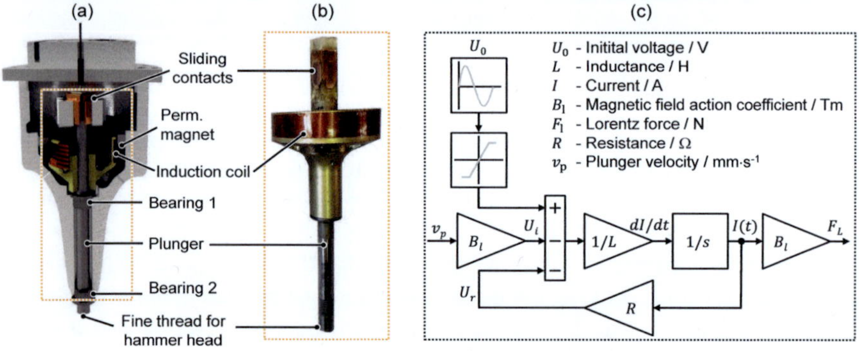

Figure 4.17: Inside view of the electrodynamic hammer tool (a), oscillating mass without hammer head in the housing of the tool (b) and diagram of the electrical domain (c)
Innenansicht des elektrodynamischen Hämmerwerkzeugs (a), schwingende Masse im Gehäuse des Werkzeugs (b) und Wirkungsplan der elektrischen Domäne (c)

The voltage supply is realized via four sliding contacts at the upper end of the oscillating mass and generates a current flow, see Figure 4.17b. In the magnetic field, the oscillating mass moves in the direction of the applied voltage. The associated force is the LORENTZ force F_L [DUEN84]. In the developed model, the electrical domain calculates the LORENTZ force F_L from the sinusoidal alternating voltage $U_0(t)$ and the returned plunger velocity $v_p(x,t)$. The diagram of the electrical domain is shown in Figure 4.17c. The voltage U_i is induced as result of the plunger movement and acts contrary to the applied input voltage U_0. Thereby, a motional electromotive force is created.

The implementation of the electrical domain is analogous to the modeling according to TRAUTH [TRAU13]. However, the real behavior of an induction coil differs from the behavior described by Equation (4.8) because of a non-linear inductance.

$$\frac{dI}{dt} = \frac{1}{L} \cdot (U_0 - R \cdot I - (\vec{B} x \vec{l}) \cdot \vec{v} \qquad \textbf{(4.8)}$$

with the initial current I, the inductance L, the initial voltage U_0, the resistance R and the plunger velocity v_p. The magnetic field coefficient B_l summarizes the expression $(\vec{B}x\vec{l})$. $\vec{B}$ is the magnetic flux density and $\vec{l}$ the length of the current-carrying wire in the induction coil.

The above-mentioned non-linear inductance results from different possible effects. Some of them are the partial skin effect [OLLE76], the proximity effect, the parasitic capacitance and especially the frequency dependence of the permeability of the magnet [KAZI13]. The skin effect is a current displacement effect in electrical inductors through which higher-frequency alternating current flows, causing the current density inside to be lower than in external areas. In general, the inductance is significantly influenced by the shape and arrangement of the windings, the wire diameter and material [REBH15]. The requirement for a linear inductance is a quasi-static magnetic field, which means that no phase differences between magnetic field and current may occur. Furthermore, a constant relative permeability number of the material in the magnetic field is necessary for a linear inductance. This number in turn depends largely on the magnetic flux density and magnetic field strength. A detailed description of the inductance in electrodynamic systems can be found in [DEMJ12] and [HERR08]. To consider a non-linear behavior of the inductor, a frequency and voltage dependent inductance $L(f, P_{el})$ was used. Therefore, the behavior of the plunger was experimentally investigated by operating the hammer system without workpiece contact at power level stages P_{el} = 10...100% in 10% steps, see Figure 4.18a and Figure 4.18b.

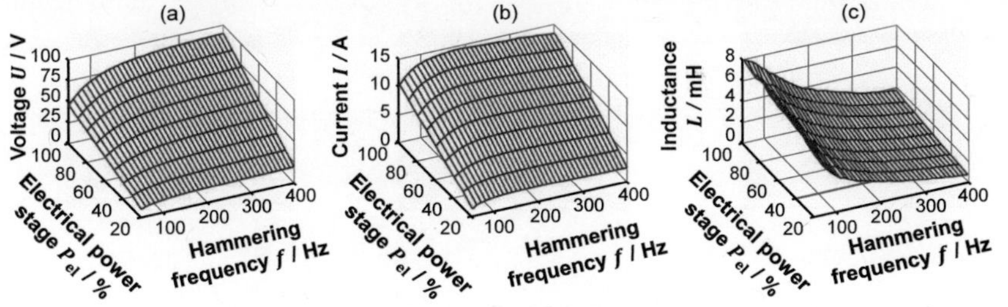

Test series: Inductance **Workpiece material**: - **MHP-parameters**: f = 50...400 Hz P_{el} = 10...100%

Figure 4.18: Dependence of voltage (a), current (b) and inductance (c) as a function of the electrical power stage and hammering frequency at MHP
Abhängigkeit der Spannung (a), des Stroms (b) und der Induktivität (c) als Funktion der elektrischen Leistungsstufe und Hämmerfrequenz beim MOH

The hammering frequencies f = 50...400 Hz were varied in 10 Hz steps. The voltage U and the current I were recorded and their amplitudes determined. Deviations of the current I due to temperature effects caused by heating or cooling phases of the hammer system were numerically smoothed. Using the interval bisection method, the inductance at which the simulated current corresponds to the experimentally determined current was determined for each data point. The derived correlation is shown in Figure 4.18c. The resistance was set to R_{mhp} = 4 Ω, close to the measured resistance in the direct current (DC) mode. The calculated frequency and power dependent inductance $L(f, P_{el})$ is capable of determining the current through the inductor for the entire operating range of electrodynamic MHP. The inductance is only slightly dependent on the power stage level P_{el}. The dependence on the hammering frequency f is more pronounced. This is due to the permeability of the permanent magnet.

As already mentioned above, input variables of the electrical domain are the supply voltage U_0 and the plunger rebound velocity v_p from the mechanical domain. The output variable of the electrical domain is the LORENTZ force F_L, which is calculated according to Equation (4.9).

$$F_L = B_l \cdot \vec{I}(t) \quad \textbf{(4.9)}$$

with $\vec{I}(t)$ the directional current.

It was assumed that the magnetic field action coefficient B_l is possibly dependent on both frequency and voltage and was therefore determined experimentally. For this purpose, the movable plunger was firmly connected with a clutch to a multi-component dynamometer 9257B from KISTLER, see Figure 4.19a.

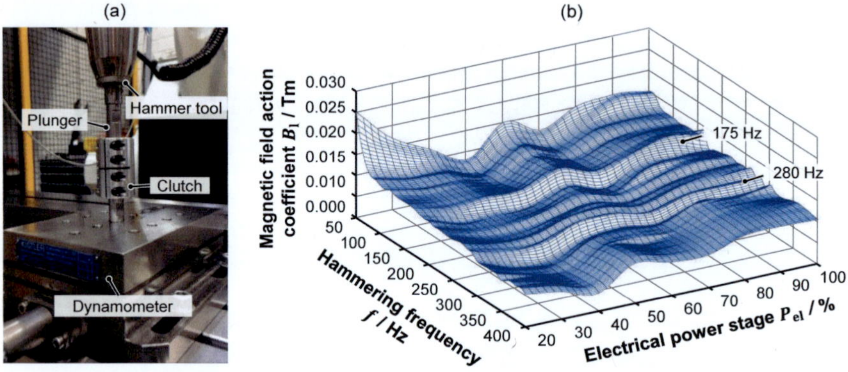

Figure 4.19: Test setup with screwed-in plunger in force measuring platform (a) and magnetic field action coefficient as a function of the electrical power stage and hammering frequency (b)

Messaufbau mit eingeschraubtem Stößel in Kraftmessplattform (a) und Magnetfeldwirkkoeffizient als Funktion der elektrischen Leistungsstufe und Hämmerfrequenz (b)

During excitation of the plunger in the hammering frequency range f = 50...400 Hz and the electrical power stage range P_{el} = 20...100%, the LORENTZ force F_L and the current I were determined. Figure 4.19b shows the calculated magnetic field coefficient B_l as a function of f and P_{el}. Obviously, the magnetic field coefficient is not constant. Nevertheless, the values fluctuate only slightly around the value B_l = 0.012 Tm. Therefore, the magnetic flux and thus the magnetic field coefficient was assumed to be constant. For the hammering frequencies f = 175 Hz and f = 280 Hz, there are excessive values for B_l. The cause are corresponding eigenmodes of the tool table. These frequencies were verified by determining a yielding frequency response from a pulse-like force signal generated by an impulse hammer. Figure A.3a in the appendix shows the spectral force in the frequency domain. The spectrum produced by a FAST FOURIER analysis shows the primary resonance peak at f = 175 Hz and the secondary resonance peak f = 280 Hz. Finally, it was examined whether the position of the plunger at a maximum free stroke h_{max} = 1.95 mm in the magnetic field of the hammer system has an influence on B_l. For this purpose, F_L was determined for three plunger positions h_p = {0.5; 1.0; 1.5} mm, see Figure A.3b in the appendix. It was found out that for P_{el} = 70% no effects of the plunger position on F_L can be deduced. For this reason, F_L is considered independent of the plunger position in the magnetic field. The magnetic field effect coefficient B_l is therefore only slightly influenced by f and P_{el}.

4.4.2 Mechanical operating system

Mechanisches Wirksystem

Within the mechanical domain, its position and velocities are determined from the forces acting on the plunger. These quantities serve as input variables for the electrical and the contact domain. The analysis of the friction forces F_r inside the hammer tool and the action diagram of the mechanical domain are shown in Figure 4.20.

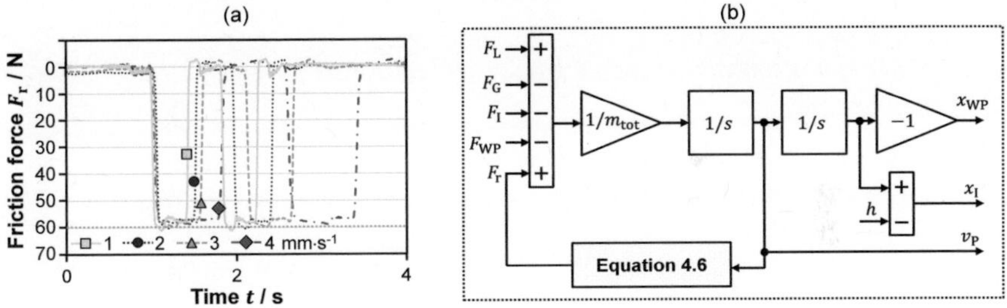

Figure 4.20: Analysis of friction forces (a) and action diagram of mechanical domain (b)
Analyse der Reibkräfte (a) und Wirkdiagramm der mechanischen Domäne (b)

The friction forces has been determined using the same test setup as in Figure 4.19a. Because of the limited free stroke, only small robot feed rates v_r = 1…4 mm·s^{-1} were analyzed. For all feed rated, a constant friction force F_r = 60 N was measured. In relation to the low hammering forces, this proportion is high and is attributed to the sliding contacts. In addition, a directional dependence of the friction force was found. The friction force influence must therefore be taken into account. Thus, the linear viscous friction was extended by a sliding component. According to Figure 4.20b, there are five forces acting on the oscillating mass. The Lorentz force F_L was already calculated in the electrical domain. The two contact forces on the inner stop F_I and on the workpiece F_{WP} result from the contact models in the following subsection. The gravitational force F_G corresponds to the product of the gravitational acceleration g = 9.81 m·s^{-2} and the oscillating mass m_{tot}. The frictional force F_r of the mechanical domain results from the velocity of the oscillating mass v_p and is calculated according to Equation (4.10). The friction parameters r_a, r_b, r_c and r_d are to be determined from experimental stroke data.

$$F_r = \begin{cases} r_a \cdot v_p + r_b \cdot v_p^2; & v_p < 0 \\ r_c \cdot v_p + r_d \cdot v_p^2; & v_p \geq 0 \end{cases} \quad \textbf{(4.10)}$$

with the plunger velocity v_p and the friction parameters $r_{a...d}$.

4.4.3 Contact mechanics

Kontaktmechanik

The contact domain is divided into two separate areas, one of which describes the workpiece contact and one the inner stop of the plunger. The modeling of the inner stop is modeled as purely elastic impact with an adjustable damping parameter c_I. A further important parameter is the coefficient of restitution e, which describes the relationship between the plunger velocity before and after the contact with the inner stop. By using the coefficient of restitution, statements about the energy conversion in the system can be made without knowledge of the material properties of the inner stop and the workpiece. The contact force at the inner stop F_I is given by Equation (4.11).

$$F_I = \begin{cases} c_I \cdot x_I \cdot {e_I}^2; & v_p < 0 \wedge x_I > 0 \\ c_I \cdot x_I; & v_p \geq 0 \vee x_I > 0 \\ 0 & x_I \leq 0 \end{cases} \quad \textbf{(4.11)}$$

with the plunger position relative to the hammer tool x_I and the coefficient of restitution at the inner stop e_I.

The coefficient of restitution can be identified by means of an analogy test (e.g. drop tower). The corresponding values for c_I can be determined by the interval bisection method. The formulation of Equation (4.11) corresponds to the HERTZ pressure. The action diagram of the plunger contact at the inner stop is shown in Figure 4.21a.

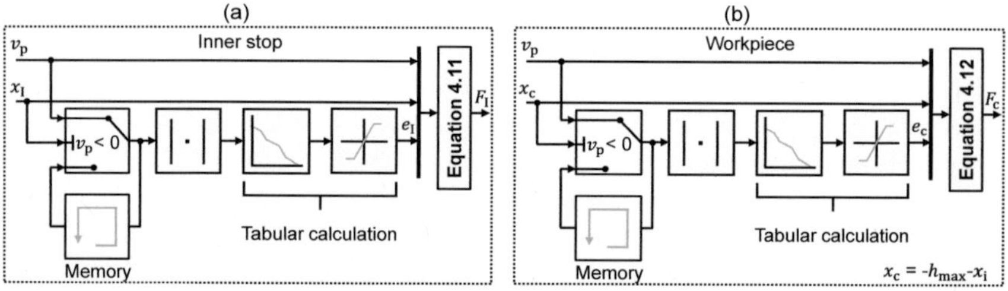

Figure 4.21: Action diagram of the contact domain at the inner stop (a) and the workpiece (b)
Wirkdiagramm der Kontaktdomäne am inneren Anschlag (a) und am Werkstück (b)

Input variables are the plunger velocity v_P and the plunger position relative to the inner stop x_I. The coefficient of restitution e_i is stored in tabular form as a function of the plunger velocity. The output value is the contact force F_i. The shown switch uses the upper input value if the condition in the middle is fulfilled; otherwise, the lower value is used. Figure 4.21b shows that the action diagram of the plunger contact with the workpiece is analogous to the action diagram of the inner stop. The equation for calculating the contact force from the coefficient of restitution at the workpiece e_{c}, the plunger position x_{WP} and velocity v_{p} uses viscous damping, which is given in Equation (4.12).

$$F_{\mathrm{wp}} = \begin{cases} f_{\mathrm{a}} \cdot x_{\mathrm{WP}}{}^{f_{\mathrm{b}}} \cdot e_{\mathrm{WP}}{}^{2}; & v_{\mathrm{p}} > 0 \wedge x_{\mathrm{WP}} > 0 \\ f_{\mathrm{a}} \cdot x_{\mathrm{WP}}{}^{f_{\mathrm{b}}}; & v_{\mathrm{p}} \leq 0 \vee x_{\mathrm{WP}} > 0 \\ 0 & x_{\mathrm{WP}} \leq 0 \end{cases} \quad \textbf{(4.12)}$$

In Equation (4.12) the plastic deformation of the workpiece is taken into account using the damping parameters f_{a} and f_{b}. The movement of the plunger can be described in both equations in the direction of the contact partner by $F = F_{\mathrm{Hertz}}$ and for the return direction by $F = F_{\mathrm{Hertz}} \cdot e^{2}$. The damping parameters can be determined using numerical optimization on experimental data from drop tower test.

4.4.4 Deduction of a data-based MHP kinetics model

Herleitung eines datenbasierten MOH-Kinetikmodells

The results presented in the previous subsections were consolidated and implemented into a fast numerical calculation model for the determination of the MHP kinetics. By applying a supervised machine learning approach in the form of the gradient method, the underlying algorithm was learned using the experimental data, see Figure 4.22.

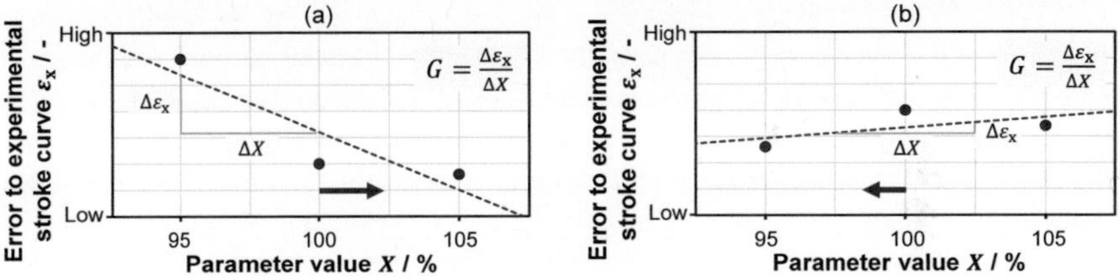

Figure 4.22: Schematic representation of the gradient method with negative (a) and positive (b) gradient

Schematische Darstellung des Gradientenverfahrens mit negativem (a) und positivem (b) Gradienten

At the beginning, the model parameters to be optimized are selected. Possible parameters are the friction and damping parameters, the coefficients of restitution and the magnetic field effect coefficient. Additional inputs are the real stroke characteristics, the learning rate and the number of repetitions with which the optimization is performed. In the first step, a recorded stroke curve is evaluated and from this the hammering frequency f, power level P_{el} and the maximum stroke h are determined. An area of the stroke curve is selected, which contains two contacts each at the upper stop and on the workpiece. With the above-mentioned input data, one simulation is initiated and a range with two contacts at each of the two stops is considered from the resulting stroke curve. The result of the simulation is both discrete in value and time and is continuously supplemented by linear interpolation. With the method of least square, the error between simulated and real stroke course is calculated and saved. For each parameter selected for calibration, two further stroke curves are then simulated in which the corresponding parameter value is varied by ± 5%. All other parameters remain unchanged during these simulations. The error between those curves and the real stroke curves is determined again. From the total of three determined errors, the parameter value change to be made is derived using the gradient method. A detailed explanation of the gradient method can be found in [KALM11]. The procedure is shown schematically in Figure 4.22.

The gradient G corresponds to the slope of a regression line through the three points. From the magnitude and sign of G, the amount and direction of the parameter value change can be derived. As indicated by the arrows, a negative gradient causes an increase of the parameter value; a positive gradient causes a decrease. Subsequently, the new parameter value is determined according to Equation (4.13). This approach is based on the work of RIEDMILLER [RIED94].

$$X = X \cdot (1 - 0.1 \cdot \tanh(G \cdot \beta_{\mathrm{L}} \cdot 10^9)) \qquad \textbf{(4.13)}$$

with X the actual parameter value, G the gradient and β_{L} the learning rate. Since the error can become very large at high gradients, the function contains the tangent hyperbolics, which is limited to (-1; 1) and the maximum correction to 10% of the initial parameter value. The learning rate gives the possibility to make the error larger or smaller within this 10%.

By means of the developed kinetics model, the system behavior can be predicted over the entire range of MHP process parameters relevant for the energy input, independent of the system generation. In addition, the numerical modeling of the process allows the investigation of materials that have not been characterized experimentally. For a new material, only the determination of the coefficient of restitution, which is explained in chapter 0, has to be performed. Subsequently, the kinetics can be predicted in good approximation. The graphical user interface of the developed MATLAB model is shown in Figure 4.23.

By means of the method of the least squares and the gradient method [GEIG99], the experimentally determined stroke courses were numerically approximated, see Figure 4.23a. This was done for the entire MHP parameter range. The calibration of the model can be made more precise and accelerated by individual selection of the system parameters. A learning rate and the number of repetitions is defined to further specify the learning process. The batch size defines the scope of the experimental measurements involved in the calibration process.

The experimental coefficient of restitution only has to be determined selectively for a new material and serve as input variable. In the further calibration, this quantity is treated like a normal system parameter. After learning the model, e.g. on a hammer system of another generation and a different material, parameter studies can be performed, see Figure 4.23b. The results are the stroke courses, contact energies and energy densities of the model and can be output later e.g. as xlxs.-file.

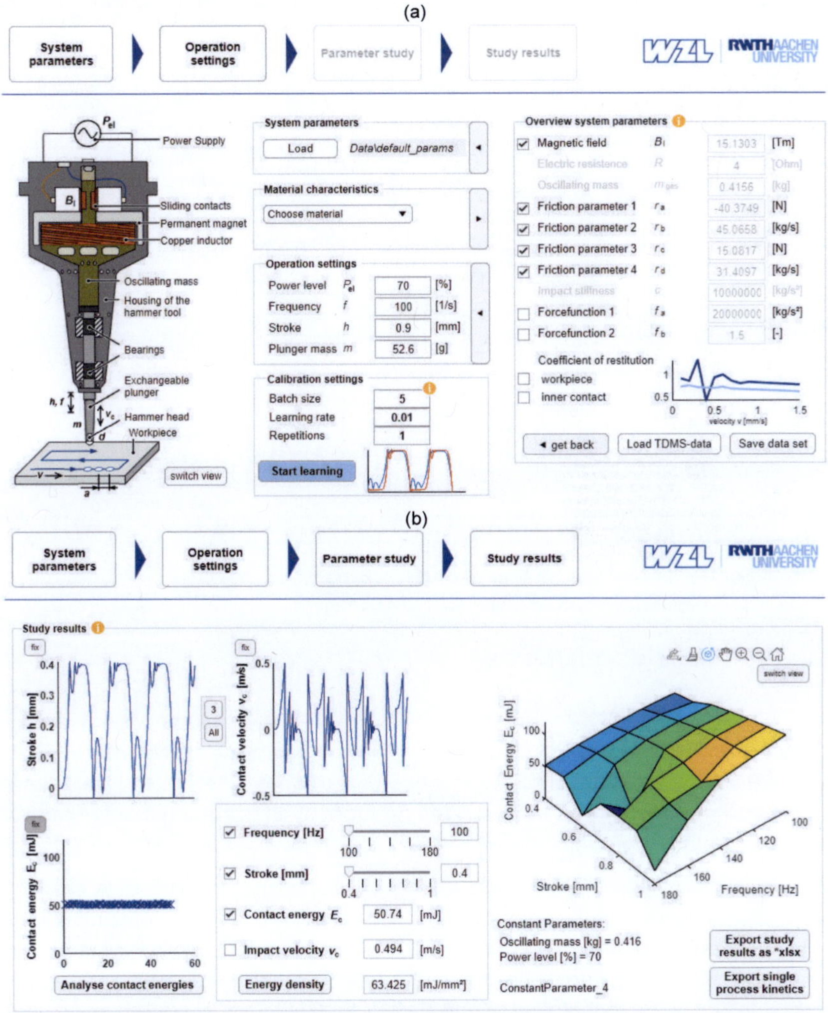

Figure 4.23: Graphical user interface of the MHP kinetics MATLAB model with the section of system parameters and parameter calibration (a) and the result display (b)
Grafische Benutzeroberfläche des MOH-Kinetik MATLAB-Modells mit dem Bereich der Systemparameter und der Parameterkalibrierung (a) sowie der Ergebnisdarstellung (b)

The developed model was validated by comparing the experimental with the numerically determined stroke courses, see Figure 4.24. For most MHP parameter settings, the stroke progression is very well mapped. The adjusted stroke value is precisely mapped numerically. Multiple impacts and irregularities in the stroke progression can be well represented with the model. Nonetheless, it must be mentioned that some of the curves still deviate from the experimental data. This results from a low learning rate and number of repetitions when learning the algorithm. An increase of the learning rate, batch size and repetitions will increase the precision. Nevertheless, the computing time required for this must be kept in mind. Furthermore, the model has to be checked for its suitability by comparison with experimental data, which do not belong to the learned parameter space.

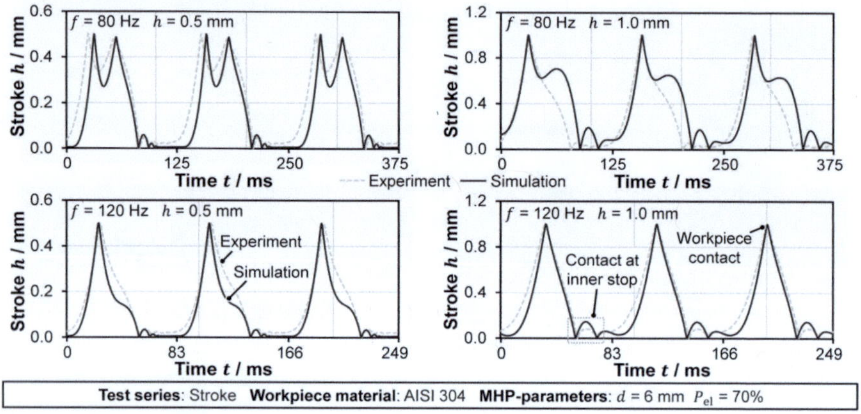

Figure 4.24: Validation of the MHP kinetics MATLAB model
Validierung des MOH-Kinetik MATLAB-Modells

4.5 Intermediate conclusion

Zwischenfazit

Chapter 4 dealt with the research question **"How do the process parameters of robot-guided electrodynamic MHP influence the process kinetics and thus the resulting contact energy and how can the multiphysical system behavior be modelled to predict the process kinetics?"**

By analyzing the influence of an industrial robot as a carrier system for the electrodynamic MHP system, it could be shown that the oscillations and elastic compressions of the hammer head occurring in the axial impact direction influence the set stroke only in the lower µm range. The deviations in the machining plane are within the order of magnitude of a single indentation on the surface and have hardly any influence on the introduced energy density.

By means of the sensor-supported extension, the voltage, current, hammer forces and plunger stroke were determined precisely inline for the first time. The analysis of the free stroke showed that the square of the frequency has a non-proportional effect on the maximum achievable stroke. Between the hammering frequency f = 100...400 Hz, the maximum free stroke changes from h_{max} = 1.95...0.08 mm. Up to a hammering frequency f = 200 Hz, the voltage and electrical power increase and remain approximately constant beyond this. The course is described by a root function. A maximum electrical power P = 713 W was determined. With increasing stroke and higher electrical power level of the MHP system, higher hammering forces are achieved up to a maximum $F_{\mathrm{h,max}}$ = 2.36 kN. With increasing stroke, the maximum hammering force shifts to lower hammering frequencies. A maximum impact velocity v_{c} = 620 mm·s^{-1} and corresponding kinetic contact energy E_{c} = 84 mJ were determined accordingly. The stroke characteristics vary greatly depending on the MHP process parameters and different contact situation on the workpiece result (e.g. multiple impacts). In the considered hammer frequency range, strokes $h \geq 0.3$ mm are characterized by a symmetrical impact profile, short contact times and thus good energy conversion conditions.

An empirical-mathematical function was developed to predict the impact velocities during MHP and the model parameters were determined using nonlinear regression to the experimental data. The model predicts the impact velocities with a mean percentage deviation of 8%. Furthermore, a numerical multiphysical process model of MHP kinetics was developed in MATLAB/SIMULINK and extended with machine learning algorithms (gradient method). The algorithms trained with experimental MHP data are able to predict the kinetics (e.g. contact energy and energy density) of the electrodynamic MHP with high accuracy. Once the correlation between the MHP parameter settings and resulting contact energies were known, **the question arose as to the effect of these contact energies on the surface integrity of individual indentations**. The answer to this question is given in Chapter 5.

5 Plastomechanical single impact analysis

Plastomechanische Einzeleindruckanalyse

Based on the results from Chapter 4 an understanding of the system-side influences on the contact conditions (contact energy E_c) in electrodynamic MHP was created and the influence of an industrial robot ABB IRB6660-205/1.9 as carrier system was evaluated. By means of the knowledge synthesized in a numerical process model, it will be possible to predict the MHP kinetics in the future. Nonetheless, the effects of defined contact energies on the geometric indentations during MHP and the resulting surface integrity of AISI 304 stainless steel are still unknown. Due to the indentations overlap that occur, the analysis of individual impacts in the real MHP process is not possible. For this reason, load analogies must be used. Therefore, first investigations of WIED by means of a drop tower [WIED11] and LECHNER by means of MHP tests with high feed rates [LECH14] on single indentation geometries of selected materials serve as starting point in the following.

In Chapter 5.1, a drop tower test bench is developed, which allows an isolated consideration of the geometric characteristics of single impacts. By analyzing different impulses and energy states on the indentation geometry, the possibility of a result transfer to a real MHP process is evaluated. Depending on defined contact conditions, an experimental analysis of separate influences on surface integrity and microstructure is carried out in Chapter 5.2. In Chapter 5.3, a numerical FE-model of the drop tower test is developed, with the help of which hard-to-measure plastomechanical quantities such as plastic strains and strain rates are analyzed. Chapter 5.4 concludes with an intermediate conclusion and answers the research question **"How do defined contact energies in the electrodynamic MHP affect the contact conditions and the geometry of individual impacts and how do these influence the surface integrity of an austenitic stainless steel?"**

5.1 Development of drop tower test rig

Entwicklung eines Fallturmprüfstandes

Due to the specific kinematic of the electrodynamic MHP process with high frequency impacts, a systematic analysis of the effects of single indentations is difficult. The approach of WIED [WIED11] to create a load analogy to the MHP by means of a drop tower test bench allows the specific investigation of surface layer effects depending on defined contact conditions (contact energies E_c). This enables the influences of different contact conditions to be considered in isolation and initial conclusions to be drawn about the mechanisms of action prevailing in MHP. In the following, the development of a drop tower test as well as a critical consideration of a result transfer to the real MHP is therefore presented.

5.1.1 Test rig design for single impact analysis

Prüfstandsgestaltung zur Einzeleinschlaganalyse

The development of an analogy test bench (drop tower) is shown in Figure 5.1a. The test stand with a total height h_{dt} = 1,000 mm consists of a base, intermediate and top plate made of C45 steel (DIN 1.0503, AISI 1045). The plates are bolted together via four aluminum profiles and angles. The base plate has a thickness t_{bp} = 40 mm. A cross table with the dimensions 225 x 175 mm was fixed via T-notches in the base plate. This enables the positioning accuracy of the clamped workpiece $\Delta x, y = \pm$ 10 µm. By clamping the workpieces in a vice on the cross table, relative movements to the base plate could be avoided. The intermediate plate, which can be positioned flexibly in height, serves to guide a hard anodized shaft made of EN AW 6060 with a diameter d_{s} = 12 mm. The shaft had a hardness 75 HB and a surface roughness Ra = 0.22 µm and is guided through a brass bushing. The ACCURAPULS plungers used for MHP were screwed together with a thread at the lower end of the shaft. To set a defined drop mass, additional metal discs were screwed on via an external thread. The shaft was held in a defined position by means of an electromagnet at the upper end for drop height adjustment.

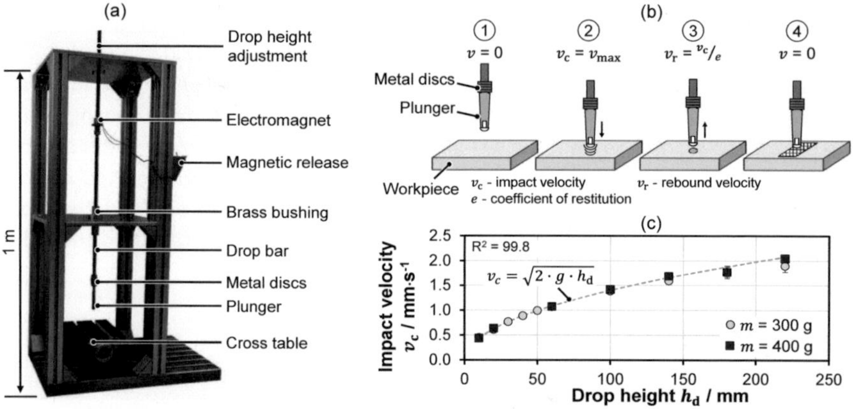

Figure 5.1: Drop tower test bench (a), four phases of plunger movement during drop tower test (b) and investigation of the influence of bearing friction and residual magnetism on impact velocity v_{c} (c)

Fallturmprüfstand (a), vier Phasen der Stößelbewegung (b) und Analyse der Lagerreibungseinflüsse und des Restmagnetismus auf die Aufprallgeschwindigkeit v_{c} (c)

Figure 5.1b shows the individual phases of the drop tower tests. At the beginning, the drop mass is held in a defined position by the electromagnet (1). After triggering the magnet, the drop mass hits the workpiece surface (2) at a maximum velocity (impact velocity v_{c}). Immediately after conversion of the kinetic energy into forming work, the plunger leaves the workpiece at an impact velocity reduced by the coefficient of restitution e (rebound velocity v_r) (3). The contact phase can be described by elastic compression, elastoplastic compression and elastic restitution. The coefficient of restitution represents the elastic part of a rotation-free impact and is therefore a measure of the energy loss. The concept and the evidence of the coefficient of restitution is used in this work and can be understood in the work of PATIL and HIGGS [PATI17]. To avoid multiple impacts, a sheet was inserted between the workpiece and the plunger (4).

Preliminary tests were carried out to analyze the influence of bearing friction and residual magnetism on the impact velocity of the plunger. Therefore, a plunger with a hammer head diameter d = 6 mm was dropped from different heights h_d = 10...220 mm with two different total masses each m = 300, 400 g. The impact velocity v_c was determined by means of a high-speed camera PHANTOM v7.3. The results are shown in Figure 5.1c. The dotted line describes the ideal conditions at constant total energy of the system. Thereby, the potential energy E_{pot} of the falling mass m is completely converted into kinetic energy E_{kin} during the drop tower tests; see Equation (5.1).

$$E_{pot} = m \cdot g \cdot h_d = \frac{1}{2} \cdot m \cdot v_c^2 = E_{kin} = E_c \tag{5.1}$$

with g = 9.81 m·s^{-2} is the gravity.

Since the measuring points lie on the dotted line with only slight deviations, a negligible influence of the bearing friction, the air resistance, the buoyancy forces and the residual magnetism of the magnet attached to the electromagnet can be assumed. This is consistent with the results of ARYAEI ET AL. [ARYA10]. Regarding the impacted workpiece, energy dissipations such as sound waves, stress waves and heat generation can be neglected for low-velocity impacts according to PATIL and HIGGS [PATI17].

5.1.2 Evaluation of result transferability to MHP process

Bewertung der Ergebnisübertragbarkeit auf MOH-Prozess

For the following drop tower tests, the energy could thus be calculated in advance on the basis of the drop mass and drop height. For all drop tower tests, milled AISI 304 specimens with the dimensions 100 x 40 x 30 mm (L x W x H) were used. The specimens were ground plane-parallel with an average roughness Sa < 1 µm. The influence of the plastification of roughness peaks on the strain states and the resulting surface integrity could thus be reduced [STIC16]. Table 5.1 shows the mean value of the chemical composition determined by spark spectroscopy out of three measurements. The requirements of DIN EN 10088-3 are met [DIN14B].

Table 5.1: Chemical composition of the used AISI 304 stainless steel
Chemische Zusammensetzung des untersuchten AISI 304 Edelstahls

Elements	C	Si	Mn	P	S	Cr	Ni	N
Weight-%	0.014	0.357	1.943	0.024	0.002	18.32	8.607	0.058

Since the drop tower test bench is merely a load analogy to the real MHP process (without tangential frictional influence), the deviations in the generated indentation geometries resulting from different impulses were also quantified. An evaluation of the indentation geometries in the drop tower is only possible at sufficiently high impulses, but with contact energies similar to MHP. For this reason, different impulses $I = I_1$, I_2 ($1.07 \cdot I_1$), I_3 ($1.15 \cdot I_1$) at selected contact energies E_c = 20...320 mJ were set by varying the drop mass m. The results are shown in Figure 5.2.

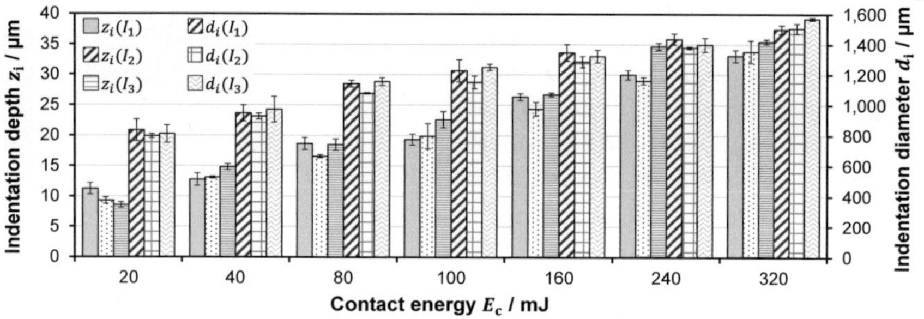

Figure 5.2: Analysis of the influence of different impulse states at selected contact energies on the resulting indentation geometry
Analyse des Einflusses unterschiedlicher Impulse bei definierten Kontaktenergien auf die resultierende Eindruckgeometrie

With increasing contact energy the indentation depth and diameter increase. Using the same impulse, the indentation geometries increase with increasing contact energy. There is no clear correlation between an increase in impulses and tendencies in modification of indentation geometries. Furthermore, the differences of indentation depths z_i and indentation diameters d_i as a result of different impulses for one specific contact energy E_c are very small. Especially the deviations regarding the indentation diameter are $\Delta d_i < 6\%$. For this reason, different impulses at similar contact energies have a negligible influence on the indentation geometry. A transferability of the geometric indentation data because of different contact energies and impulses from the drop tower to the real MHP process is therefore possible.

5.2 Experimental analysis of indentation geometry and contact modes

Experimentelle Analyse von Eindruckgeometrie und Kontaktmechanismen

After the development and qualification of a drop tower test rig for the analysis of the effects of defined contact energies from individual impacts on the surface integrity, experimental tests were carried out. In addition to the analysis of the impact and rebound velocities and thus the energy balance in the contact between hammer head and workpiece, the geometric characteristics of the indentations were also investigated.

5.2.1 Procedure and evaluation of indentations

Versuchsdurchführung und Eindruckauswertung

The single indentation tests were performed on the drop tower test rig developed in Chapter 5.1 and by means of the electrodynamic MHP system described in Chapter 4. Three test series were defined for the analysis of single and overlapping indentations.

Drop tower tests for single indentations

In this test series, the undisturbed influence of individual indentations on the workpiece surface layer was investigated. Drop tower tests with constant total mass m = 375 g were carried out to analyze impact and rebound velocities as well as surface integrity (indentation geometry and microstructure) of single indentations. According to JOHNSON, the impact velocity, the drop mass, the hammer head diameter and the material properties have a certain influence on the remaining indentation geometry when a ball hits a plane surface [JOHN85]. Since the oscillating mass is approximately constant in the real MHP process, only the impact velocity and the hammer head diameter were varied. The hammer head diameter d = {6; 8; 10; 12} mm as well as the drop height h_{d} = 10...220 mm were changed to set different impact velocities and thus resulting contact energies and surface pressures.

Drop tower tests for overlapping indentations

Since in real MHP, individual indentations overlap and thus influence the immediate surroundings of one single indentation, drop tower tests were also carried out using a cross table. The cross table allows for precise positioning of the specimen under the falling mass. The test series was also executed with a constant total mass m = 375 g. A hammer head diameter d = 6 mm was selected. The indentation distances a = {0.05; 0.2; 0.5} mm and stepover distances s = {0.2; 0.5} mm were varied. For the overlapping tests in both plane directions, $a = s$ was chosen.

MHP tests for single indentations

Two types of MHP experiments to generate individual indentations were selected. On the one hand, MHP tests during defined times t_{mhp} = {2; 10} s on one point with set stroke h = {0.1; 0.5; 1.1} mm were executed. In terms of the affected area in the microstructure, LIDA found no significant difference from a single impact [LIDA00]. On the other hand, MHP tests with constant stroke h = 0.5 mm and varied feed rates for setting defined indentation distances a = {0.05; 0.5; 1.0} mm were performed. A hammer head diameter d = 6 mm was used for all experiments. By choosing a small hammerhead diameter and thus higher contact pressure, effects become more apparent.

Analysis methods

The impact and rebound velocities were determined by means of the high-speed camera from Chapter 4, see Figure 5.3a. To determine the impact velocities, ten measurement points (orange points) each were evaluated before and after the impact.

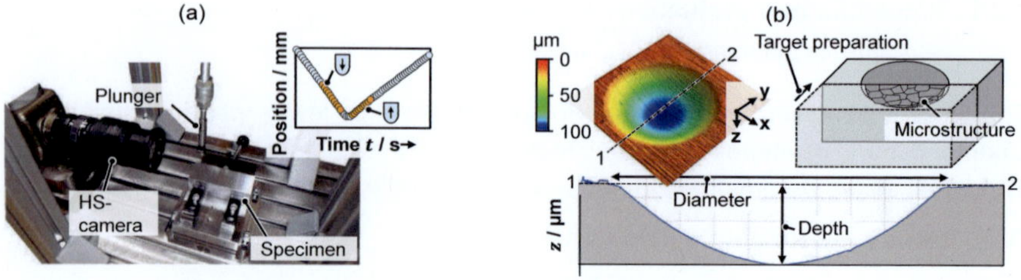

Figure 5.3: Determination of the plunger position using a high-speed camera (a) and topography and microstructure analysis of a single indentation (b)
Ermittlung der Stößelposition mittels Hochgeschwindigkeitskamera (a) und Topografie- und Mikrostrukturanalyse eines einzelnen Eindrucks (b)

The dimensions of the indentations (depth z_i, diameter d_i and volume V_i) were determined by means of a measuring system Hommel Etamic T8000 RC according to DIN EN ISO 25178 [DIN12], see Figure 5.3b. The microstructure was made visible using a target preparation by means of AccuStop-T specimen holder of STRUERS GMBH. Firstly, the individual indentations were separated under low heat influence by means of a precision cutting machine STRUERS Discotom-5. The material was cut off up to approximately 500 µm from the indentation edges. By means of grinding papers of different grit 120...4000, the indentation was carefully grinded manually on a grinding machine BUEHLER PowerPro 4000 to within approximately 20 µm of the indentation. Between the individual grinding steps, a light microscope ZEISS Axio Imager M2m was used to analyze the remaining distance to the center of the indentation. After the grinding steps, the samples were then polished with a suspension of 1 µ to produce a deformation-free surface. BERAHA II and V2A were used as etching solutions.

5.2.2 Results of contact conditions and indentation morphology

Ergebnisse der Kontaktbedingungen und Eindruckmorphologie

In the following the results of the above mentioned three test series are discussed. Thereby, the effects of different contact energies on the resulting indentation geometries and especially the microstructure are analyzed. The contact energy resulting from the impact velocity is converted during the impact into elastic and plastic deformation energy as well as frictional heat. Since a rotation-free central impact is considered here, the loss energy converted into heat by friction (at v_c < 10 m·s^{-1}) can be neglected compared to the plastic deformation energy according to TABOR [TABO00].

Figure 5.4 shows the results of the impact velocities v_c and rebound velocities v_r for different hammer head diameters d. The coefficient of restitution e is the quotient of v_r to v_c. The determination of the coefficient of restitution enables statements to be made about the energy state of the overall system.

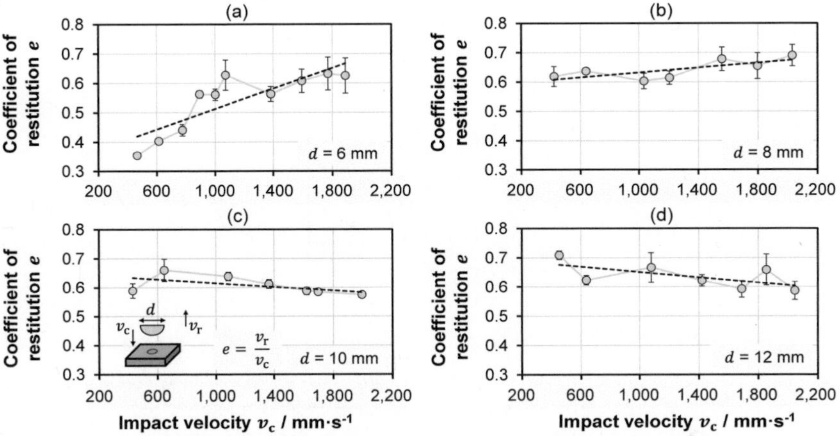

Figure 5.4: Experimental results of coefficient of restitution e as function of impact velocities v_c for different hammer head diameters d = 6 (a); 8 (b); 10 (c) and 12 mm (d)
Experimentelle Ergebnisse der Stoßzahl e als Funktion der Aufprallgeschwindigkeiten v_c für unterschiedliche Hammerkopfdurchmesser d = 6 (a); 8 (b); 10 (c) und 12 mm (d)

A certain scattering of the determined coefficient of restitution e can be observed. The reason for this may be microscopic surface roughness of the two contact partners. This results in an energy transfer between the translational and rotational degrees of freedom [MONT11]. For most of the investigated combinations of impact velocity and hammer head diameter, the coefficient of restitution is between e = 0.6...0.7. This coincides with the observations of SANDEEP ET AL. on stainless steel [SAND20]. For the hammer head diameters d = {6; 8} mm, the coefficient of restitution increases with increasing impact velocity. This contradict the results of BACA ET AL. [BACA16]. A possible reason may have been inadequate specimen clamping and thus springing of the specimen when it was struck. SONDERGARD ET AL. describe this phenomenon by the occurrence of shear and RAYLEIGH surface waves resulting from the clamping [SOND90].

It is also conceivable that in the case of the smaller hammer head diameters, the higher contact pressure leads to a conversion of austenite into harder martensite. This effect is intensified by the work hardening occurring in contact due to increased dislocation densities. Because of the higher hardness of the martensite phase and the increased dislocation density, the elastic deformation portion of the workpiece increases. Thus, the rebound velocity v_r of the plunger and therefore the coefficient of restitution increases. For the larger hammer head diameters d = {10; 12} mm, e decreases with increasing v_c. It is also noticeable that the effect of an increasing or decreasing impact velocity is more pronounced for small hammer head diameters d = 6 mm and large hammer head diameters d = 12 mm.

The results of the indentation geometries are shown in Figure 5.5. The indentation diameter d_i and the indentation depth z_i increase with increasing contact energy E_c.

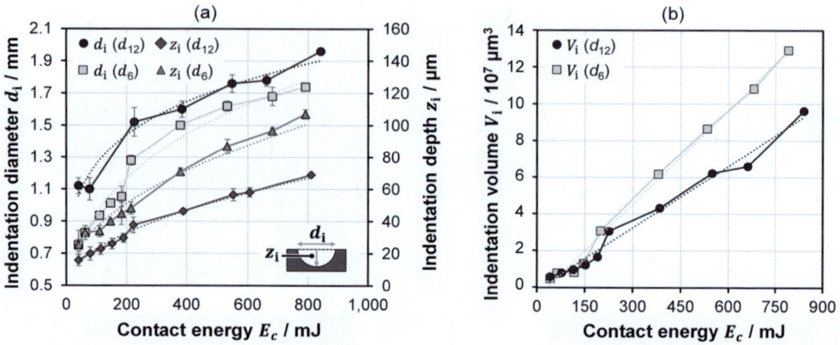

Figure 5.5: Results of indentation diameter and depth (a) as well as indentation volume (b) for selected contact energies and hammer head diameters
Ergebnisse der Eindruckdurchmesser und -tiefe (a) sowie Eindruckvolumen (b) für ausgewählte Kontaktenergien und Hammerkopfdurchmesser

The relation between these two geometric indentation characteristics can be described by means of a power function. KANOU ET AL. determined a similar relationship for higher impact velocities and higher contact energies [KANO12]. Differences are due to the higher strain rates resulting at higher impact velocities. Thereby, the yield stress increases, the material become less plastically deformable and the indentation geometry differs. With a larger hammer head diameter d = 12 mm, great indentation diameters but low indentation depths are achieved compared to a small hammer head diameter d = 6 mm. With a larger hammer head, the introduced contact energy is distributed over a greater contact surface, resulting in greater material displacement in the tangential plane direction. Since the contact energy has been kept constant, the ratio between indentation diameter and indentation depth is higher for the same indentation volume with a larger hammer head diameter. For both hammer head diameters an almost linear increase of the indentation dimensions with increasing contact energy is found. The results confirm the findings of WIED [WIED11A] and GOLDSMITH [GOLD15].

Provided that the elastic deformation of the material is negligible, these test results also confirm the linear relationship between the indentation diameter and the impact velocity according to JOHNSON [JOHN85]. Regarding the indentation volume V_i, it is noticeable that with increasing contact energy, the displaced material volume increases more with a smaller hammer head diameter.

The results of the microstructure of single indentations with a hammer head diameter d = 6 mm are shown in Figure 5.6. To visualize the microstructure, images of samples etched with BERAHA II-solution were taken.

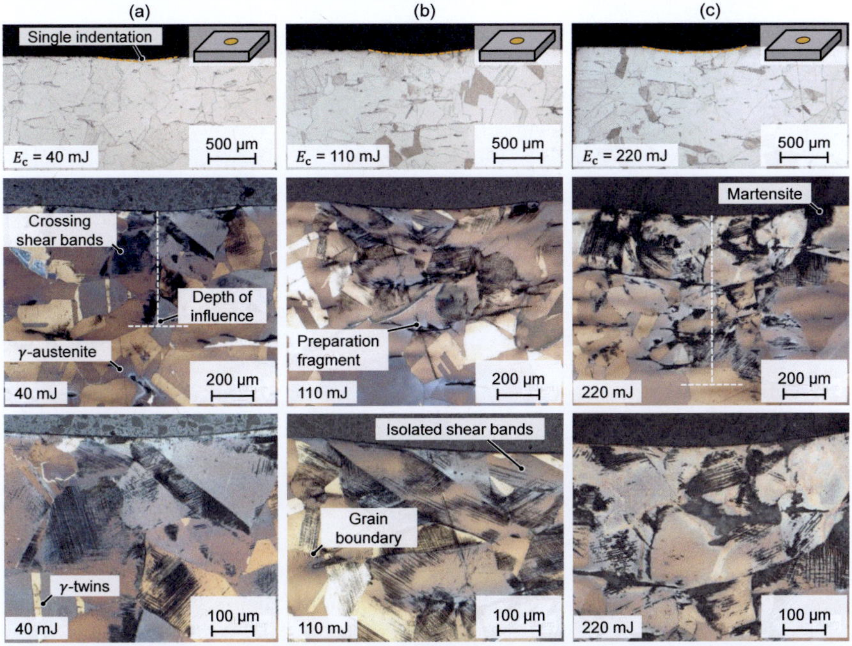

Figure 5.6: Microstructural analysis of areas below single indentations produced with a hammer head diameter d = 6 mm and different contact energies E_c = 40 mJ (a), 110 mJ (b) and 220 mJ (c)

Mikrostrukturanalyse von Bereichen unterhalb von Einzeleindrücken, die mit einem Hammerkopfdurchmesser d = 6 mm und unterschiedlichen Kontaktenergien E_c = 40 mJ (a), 110 mJ (b) und 220 mJ (c) erzeugt wurden

For all three contact energies E_c = 40; 110; 220 mJ, dark areas have formed below the indentations. These represent shear bands as well as martensite, which was shown by means of micro hardness measurements in Figure 5.7a. A clear assignment to the α'-martensite or ε-martensite cannot be made. It is assumed that the proportion of ε-martensite predominates due to the only slight increase in hardness. Isolated shear bands consist of ε-martensite and twins and appear in only one direction. With stronger accumulation and crossing of the shear bands α'-martensite can form. The blue and brown areas belong to the initial γ-austenitic phase [MÜLL11]. An increased accumulation of the dark areas appears at the edges of the indentations. With increasing contact energy E_c, the dark areas become more pronounced and denser. While the change in microstructure at low contact energy still reaches a depth of z = 550 µm below the indentation, the influenced depth increases to z = 880 µm at higher contact energy. This increase is discontinuous.

The micro hardness measurement showed good agreement with the depth of the dark areas. The graphs in Figure 5.7b show the hardness of the initial material state and under an indentation created with E_c = 220 mJ.

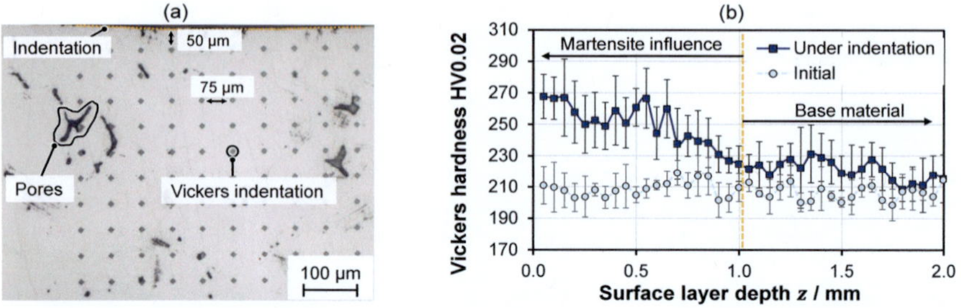

Figure 5.7: Micro hardness measurement grid in polished cross section (a) and micro hardness depth profile (b)
Mikrohärtemessraster in poliertem Querschliff (a) und Mikrohärte-Tiefenverlauf (b)

A maximum hardness increase of $\Delta HV0.02$ = 30% could be measured. Taking into account the scatter of the values, the initial microstructure is reached from approx. 1 mm surface layer depth. The inhomogeneous microstructure change confirms the assumption that the kinetics during ball impact is influenced by phase transformations.

For better visualization of the influenced areas underneath the indentations, these were polished again for the single and overlap tests after the BERAHA II etching, see Figure 5.8. An impairment of the deformation zone could be excluded by comparing them with unpolished samples, see Figure A.5 in the appendix. The zone of action in the microstructure is many times larger than the plastic indentations. A similar observation was made by LIDA on a carbon steel using FRYS etching method [LIDA00].

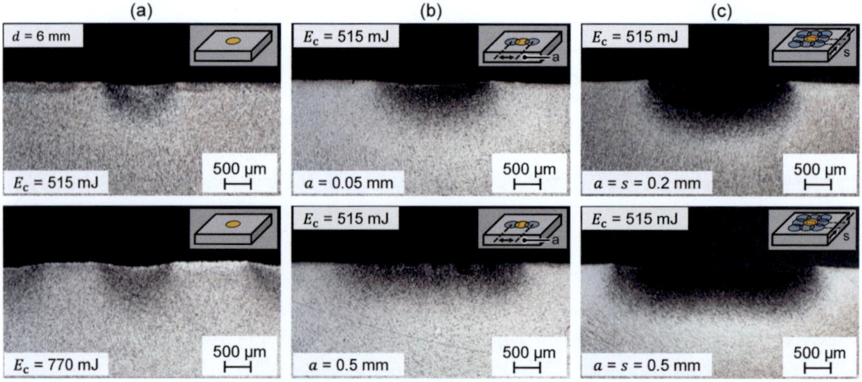

Figure 5.8: Influence of contact energy (a), indentation distance (b) and indentation density (c) on microstructure with a hammer head diameter d = 6 mm
Einfluss der Kontaktenergie (a), des Eindruckabstandes (b) und der Einschlagdichte (c) auf die Mikrostruktur mit einem Hammerkopfdurchmesser d = 6 mm

By means of the higher contact energy E_c = 770 mJ, a less pronounced dark area can be observed. One possible explanation are dislocation annihilations due to the exceeding of a critical energy threshold. Nonetheless, the depth of the affected zone below the indentations is about 850 µm for both contact energies. The lower contact energy has been used for the analysis of the line (Figure 5.8b) and the area overlap tests (Figure 5.8c). For line analysis, 9 indentations and for area tests, 9x9 indentations has been created in drop tower test.

Using smaller indentation distance leads to a more homogenous microstructure underneath the indentations. At a larger indentation distance, unconverted bright microstructure areas are visible. Thus, there is no complete overlapping of the individual stress fields. The effective depth of action remains unaffected, see Figure 5.8b. A dense, homogeneous microstructure is created by a distribution of overlapping indentations in two spatial directions. This can be explained by means of the model according to TRAUTH [TRAU16A]. During MHP in two spatial directions, individual volume elements undergo deformation in two directions. This activates additional sliding systems in the workpiece, the dislocation density increases and larger austenitic areas fold over into the martensitic phase. Figure 5.8c shows that the depth of action of the influenced zone also extends further into the material to around 1 mm. Thus, it is concluded that a sufficient single impact energy and overlap as well as the resulting energy density significantly influence the transformation microstructure of the metastable AISI 304 stainless steel.

MHP tests were performed to provide a comparison with the microstructure below the indentations created in the drop tower test, see Figure 5.9. The microstructure after MHP test on one point during five seconds with varying strokes resulting in different contact energies E_c = 3.6; 24; 51 J are shown in Figure 5.9a.

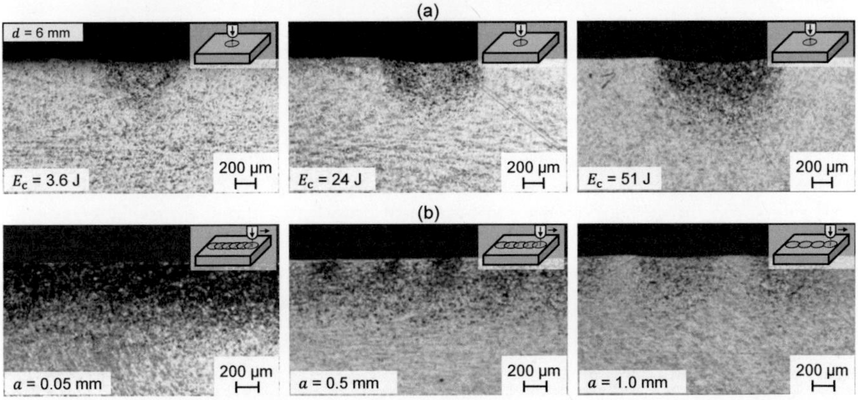

Figure 5.9: Influence of selected contact energies resulting from MHP point hammering with different strokes during five seconds (a) and evolution of microstructure as function of the indentation distance (b)

Einfluss ausgewählter Kontaktenergien beim MOH-Punkthämmern während fünf Sekunden (a) und Mikrostrukturentwicklung in Abhängigkeit der Eindruckabstände (b)

The indentation geometry increases in diameter and depth direction with increasing contact energy. Meanwhile, the depth of the zone of action slightly increases while the volume increases significantly. Higher indentation distances a during MHP lead to lower microstructural changes, see Figure 5.9b. Due to the increased feed rate, the normal force component decreases and the forces in tangential direction increase. Even with a certain overlap, a homogenous microstructure cannot be guaranteed.

Since a higher indentation distance correlates with a higher feed rate, a preferential direction of the resulting microstructure occurs. As LECHNER [LECH14] showed that MHP at high feed rates has no influence on the circular indentation geometry, the influence on the microstructure must be evaluated separately. Furthermore, it was found that the different contact conditions between the MHP and drop tower lead to differently pronounced surface layer. Because of the influenced zone, however, it can be seen that in general, findings are transferable and relative changes can be evaluated.

The cause-and-effect relationships between selected contact energies of individual and one- and two-dimensional overlapping indentations and the resulting indentation geometry and microstructure development in AISI 304 were determined. However, the strain states and strain rates in the workpiece surface layer resulting from the contact energies are still unknown. Therefore, numerical FE analysis of impact tests are presented in the following subsection.

5.3 Numerical FE modeling of impact tests

Numerische FE-Modellierung von Aufprallversuchen

The impact tests were modelled using the finite element (FE) method. The challenge in this subsection was the high mesh fineness required due to the small indentation sizes and the implementation of the experimentally measured impact velocities. Furthermore, the question had to be answered to what extent the elastoplastic material parameters can be determined and correlated to the resulting microstructure.

5.3.1 Set-up and verification of the FE model

Aufbau und Verifizierung des FE-Modells

Part modelling. In addition to the experimental drop tower tests, an explicit three-dimensional FE model of the impact tests was developed using the ABAQUS 6.14 software. The hammer head was modelled as a rigid body with a diameter d = {6; 8; 10; 12} mm. The cuboid workpiece was modelled as a deformable body with a height $h_{\mathrm{WP,FE}}$ = 5 mm and a square base area with a length $l_{\mathrm{WP,FE}}$ = 6 mm and a width $w_{\mathrm{WP,FE}}$ = 6 mm, see Figure 5.10a.

Mesh discretization. The workpiece was meshed with hexagonal continuum elements with reduced integration (C3D8R, $32.7 \cdot 10^6$ elements) with an element size l_e = 10 µm. The hammer head was meshed with rigid elements (R3D4/R3D3, $254 \cdot 10^3$ elements) with an element size l_e = 5 µm, see Figure 5.10b. These tetrahedron elements are characterized by a faster computation time. The element size was selected after performing a convergence analysis on workpiece width and depth, see Figure 5.11b.

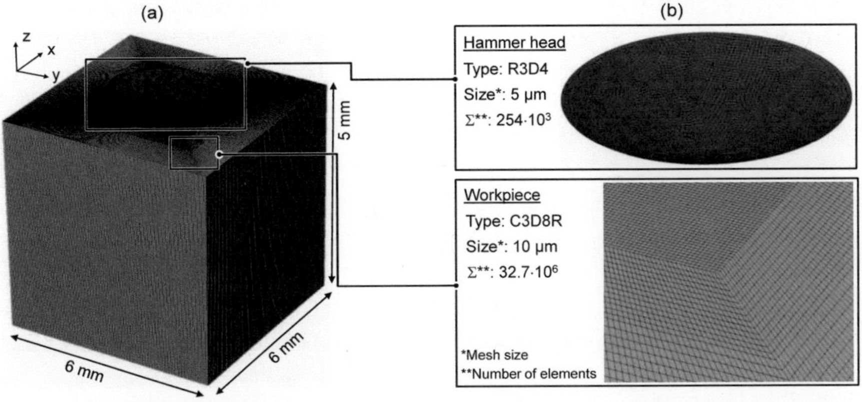

Figure 5.10: Set-up of the impact test FE model (a) and chosen element types R3D4 for the hammer head and C3D8R for the workpiece (b)
Aufbau des FE-Modells der Einschlagversuche (a) und ausgewählte Elementtypen R3D4 für den Hammerkopf sowie C3D8R für das Werkstück (b)

Material properties. The elastic properties of the AISI 304 workpiece were defined with a YOUNG'S modulus E = 200·10³ N/mm² and a POISSON'S ratio ν = 0.3. The workpiece density was defined as ρ = 7.9·10⁻⁶ kg/mm³. The plastic material behavior was modeled using a strain rate dependent JOHNSON-COOK plasticity model, see Equation (5.2).

$$\sigma = \left[A_{JC} + B_{JC} \cdot (\varepsilon_{pl})^{n_{JC}}\right] \cdot [1 + C_{JC} \cdot \ln\left(\frac{\dot{\varepsilon}_{pl}}{\dot{\varepsilon}_{pl,0}}\right)] \quad \textbf{(5.2)}$$

The model coefficients A_{JC} = 510 N/mm², B_{JC} = 703.4 N/mm², C_{JC} = 0.0175 and n_{JC} = 0.3308 were determined from experimental cylindrical upsetting tests using Curve Fitting Toolbox V3.4 from MATLAB R2019b. The upsetting tests were carried out at room temperature and three forming velocities $\dot{\varphi}$ = {1; 10; 100} s⁻¹ on a hydraulic press SERVOTEST DCS 2000 at the Institute of Metal Forming (IBF) at RWTH Aachen University. The flow curves are shown in Figure 5.11a.

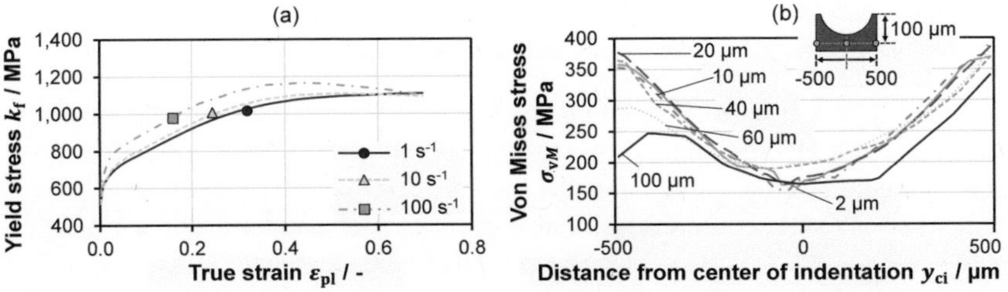

Figure 5.11: Flow curves for AISI 304 from cylindrical upsetting tests at different forming velocities (a) and convergence analysis for determination of mesh size (b)
Fließkurven für AISI 304 aus Stauchversuchen bei verschiedenen Umformgeschwindigkeiten (a) und Konvergenzanalyse zur Bestimmung der Netzfeinheit (b)

Boundary and contact conditions. For the lateral faces of the workpiece symmetrical boundary conditions, XSYMM in the positive and negative x-direction and YSYMM in the positive and negative y-direction were chosen. The bottom of the workpiece was fixed using the *encastre* option. To minimize the penetration of points on the deforming workpiece surface into the rigid MHP head surface, the normal contact condition *hard contact* was defined. This significantly minimizes over-closure of contact nodes.

Process kinematics. The mass of the falling body was defined as m = 375 g. For different contact energies E_c = 10...1,000 mJ, the impact velocities were calculated by means of equation (4.6).

Verification. The FE model was verified for the hammer head diameters d = {6; 12} mm. Therefore, Figure 5.12 shows the comparison of the numerically and experimentally determined indentation depths (a) and diameters (b). In Figure 5.12a, the indentation depths show a very good agreement. The deviations were below 6% over the energy range considered. The deviations between the experimental and simulated results of the indentation diameters are slightly higher with a maximum of 8%. Nonetheless, the FE model was assumed to be verified in order to provide reliable results for the predominant plastomechanical states below single indentations.

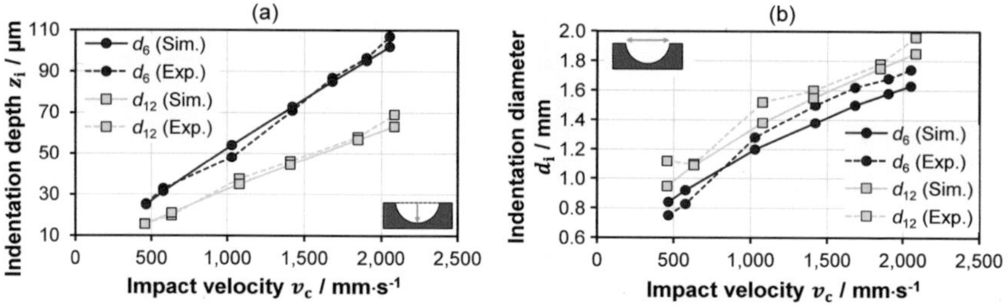

Figure 5.12: Comparison of the experimental (Exp.) and simulated (Sim.) indentation depths (a) and indentation diameters (b) to verify the drop tower FE model
Vergleich der experimentellen (Exp.) und simulierten (Sim.) Eindrucktiefen (a) und Eindruckdurchmesser (b) zur Verifizierung des FE-Modells der Aufprallversuche

5.3.2 Analysis of plastomechanical material state

Analyse des plastomechanischen Werkstoffzustandes

It is known from the state of the art that the plastic strains and strain rates resulting in the workpiece from the forming work, i.e. the contact energy, have a significant influence on the microstructure and thus the mechanical workpiece properties [CHAN10]. Furthermore, the influenced microstructural area is much greater than the indentation geometry [LIDA00]. For this reason, an analysis of the strains, the influenced zone of action and strain rates in the surface layer allows conclusions to be drawn about predominant microstructural mechanisms. Figure 5.13a shows that the areas of highest strains ε_{pl} and strain rates $\dot{\varepsilon}_{pl}$ in the FE simulations correspond to the darkest areas in the microstructural images. The occurrence of the highest strains at the indentation edges confirms the results of HU ET AL. [HU17].

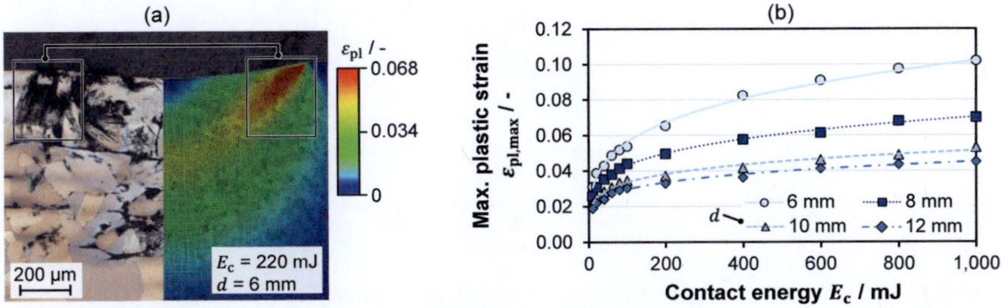

Figure 5.13: Comparison of local plastic strains with microstructural features (a) and influence of contact energy on the maximum plastic strains (b)
Vergleich der lokalen plastischen Dehnungen mit Mikrostrukturmerkmalen (a) und Einfluss der Kontaktenergie auf die maximalen plastischen Dehnungen (b)

With increasing contact energy at constant hammer head diameter, the maximum plastic strains in the surface layer increase, see Figure 5.13b. In the contact energy range up to E_c = 1,000 mJ, maximum plastic strains $\varepsilon_{pl,max} \approx 0.1$ were determined. In the MHP-relevant range up to approximately E_c = 100 mJ, maximum plastic strains $\varepsilon_{pl,max}$ = 0.021...0.054 occur. The dependence of the maximum plastic strain on the contact energy introduced can be described using the power law, see Equation (5.3).

$$\varepsilon_{pl,max} = A_{PL} \cdot E_c^{B_{PL}} \quad \textbf{(5.3)}$$

The coefficient of determination R^2 for all graphs is above 99%. The values of the parameters of the power law are listed in Table 5.2.

Table 5.2: Parameters of the power law for the dependence of the maximum plastic strain on the contact energy
Parameter des Potenzgesetzes zur Abhängigkeit der maximalen plastischen Dehnung von der Kontaktenergie

Hammer head diameter d / mm	**6**	**8**	**10**	**12**
A_{PL} / -	0.0169	0.0161	0.0144	0.0126
B_{PL} / -	0.2605	0.2132	0.1842	0.1838
R^2 / %	99.5	99.7	99.2	99.0

Furthermore, the FE simulations were used to analyze the zone of action below the indentations. The dimensions of the zone of action were also set in relation to the experimentally determined indentation geometries. Figure 5.14 shows that the dimensions of the zone influenced in the material structure by an impact are significantly larger than the geometrical indentation dimensions.

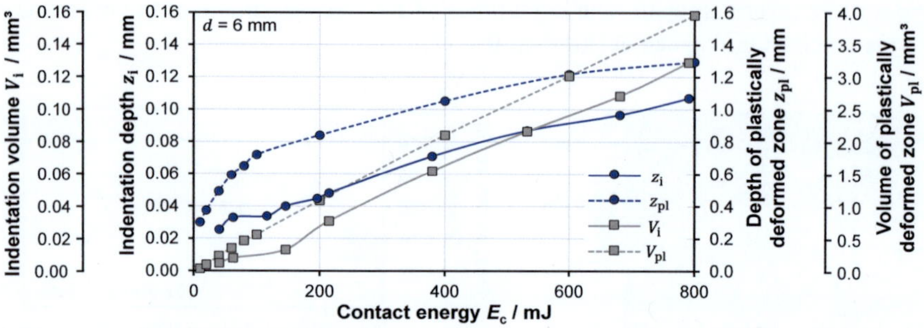

Figure 5.14: Influence of contact energy on depth and volume of deformed zone below the indentation and indentation geometry
Analyse des Einflusses der Kontaktenergie auf die Tiefe und das Volumen der plastisch deformierten Zone unterhalb eines Eindrucks und die Eindruckgeometrie

The depth of the affected zone is greater than the indentation depth by a factor of 8 on average. With increasing contact energy, the increase of z_i and z_{pl} can be described with the power law. The volume increase linearly with increasing contact energy. The influenced volume in the microstructure is higher by a factor of 30 than the resulting indentation volume. A similar behavior was observed for larger hammer heads. Figure 5.15a shows the time-varying strain rates at a surface layer depth z = 25 µm for different contact energies E_c.

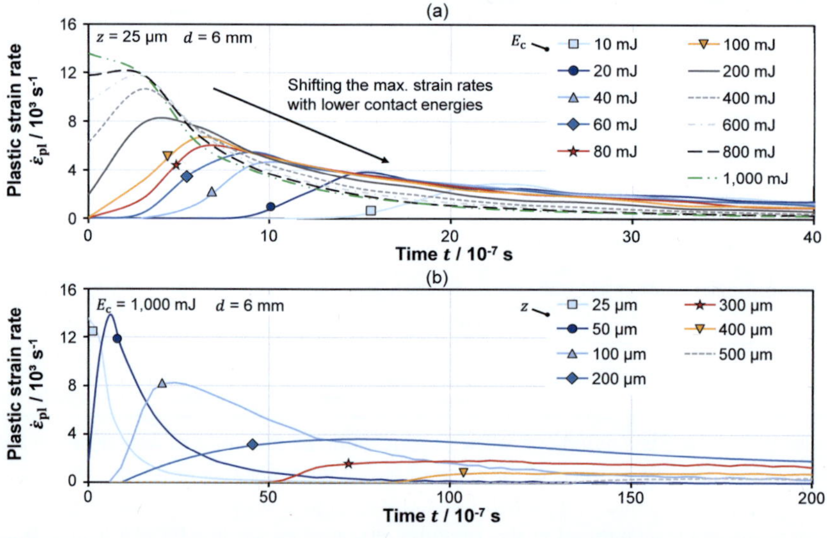

Figure 5.15: Analysis of plastic strain rates at different contact energies E_c for a surface layer depth z = 25 µm (a) and at different z for E_c = 1,000 mJ (b)
Analyse der plastischen Dehnraten bei unterschiedlichen Kontaktenergien E_c in einer Randzonentiefe z = 25 µm (a) und bei unterschiedlichen z für E_c = 1.000 mJ (b)

Figure 5.15b shows the time-varying strain rates at different surface layer depths z = 25...500 µm for a high contact energy E_c = 1,000 mJ. The plastic strain rates in the first contact moment increase with increasing contact energy. Maximum plastic strain rates $\dot{\varepsilon}_{pl,max}$ = 2,800...13,540 s^{-1} were determined for selected contact energies E_c = 10...1,000 mJ. The maximum of the plastic strain rates occurs for decreasing contact energies at later times. The plastic strain rate decreases only slightly at the beginning with increasing surface layer depth. Afterwards, it decreases sharply. STRONGE described the same phenomena for stress distribution of rigid body contact [STRO00]. In surface layer depths z > 800 µm, the plastic strain rates are already close to zero.

By means of the hypothesis of CHEN ET AL., different deformation mechanisms are to be expected according to the different strain rates. At strain rates $\dot{\varepsilon}_{pl}$ = 10^4...10^5 s^{-1}, a direct transformation into α'-martensite and the formation of deformation twins is expected. At lower strain rates up to $\dot{\varepsilon}_{pl}$ < 10^3 s^{-1}, on the other hand, dislocation developments in the form of sub-cell structures as well as disordered dislocation arrangement and α'-martensite formation will occur [CHEN11B]. As a result, single impacts of MHP are assumed to evoke a combination of martensite transformation and sub-cell structures resulting from increased dislocation densities. It could be found out that the depth of the dark areas underneath one single indentations from the experiments correlates well with the simulation results with $\dot{\varepsilon}_{pl}$ > 10^1 s^{-1}.The MHP uses different hammer head diameters depending on the contours to be hammered. This results in different surface pressures. Their influence on the plastic strain rate is shown in Figure 5.16a.

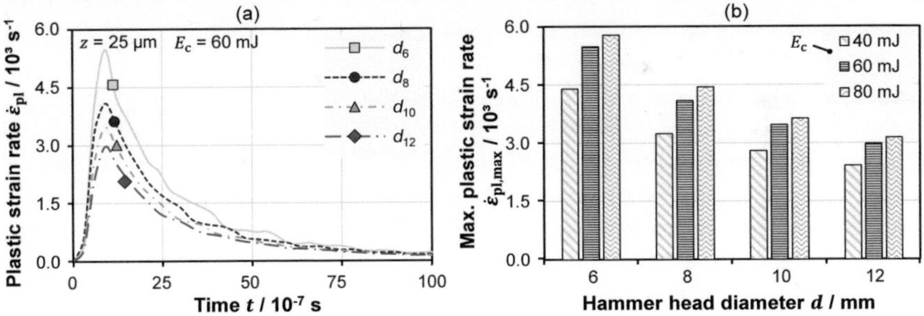

Figure 5.16: Influence of hammer head diameter d on the plastic strain rate $\dot{\varepsilon}_{pl}$ (a) and analysis of the combined influence of d and contact energy E_c on $\dot{\varepsilon}_{pl}$ (b)

Einfluss des Hammerkopfdurchmessers d auf die plastische Dehnrate $\dot{\varepsilon}_{pl}$ (a) und Analyse des kombinierten Einflusses von d und der Kontaktenergie E_c auf $\dot{\varepsilon}_{pl}$ (b)

Smaller hammer head diameters lead to an increased maximum plastic strain rate. This can be explained by the higher contact pressure [LECH14]. The analyses shown in Figure 5.16b illustrate that just by selecting a small hammer head, a near doubling of the maximum plastic strain rate can be achieved with the same energy input. It could be shown that, in addition to the contact energy, the surface pressure also has a significant influence on the plastic strain rates and thus on the microstructure.

5.4 Intermediate conclusion

Zwischenfazit

Chapter 5 dealt with the question **"How do defined contact energies in the electrodynamic MHP affect the contact conditions and the geometry of individual impacts and how do these influence the surface integrity of an austenitic stainless steel?"**

At the beginning of this chapter, it was shown that a drop tower test rig, despite neglecting the tangential contact component as well as deviating impulse conditions, is suitable for the analysis of the interactions between selected parameters (HERTZ compression, contact energy) and the plastomechanical contact conditions in MHP. This confirms the analyses of WIED [WIED11A].

By means of the developed and validated drop tower test rig, it was shown that the coefficient of restitution correlating with the energy balance of an impact shows a partially deviating behavior from the findings described in the literature [GOLD15]. In addition to clamping effects of the samples, the work hardening and transformation behavior of the analyzed AISI 304 steel was thereby discussed. With higher contact energies, an increase in indentation diameter, depth and volume was observed. The dependence of the indentation depth and diameter on the contact energy can be described by means of a power law. The indentation volume increases linearly with increasing contact energy. BERAHA II etching solution was also used to visualize the microstructural mechanisms of action below individual impacts. In addition to shear bands, martensite was generated from the originally metastable γ-austenite structure. An increased microhardness of the transformed microstructure confirms this statement. An increased expression of these dark areas in metallographic sections was detected at the edges of the indentations. The depth of the influenced microstructure layer increases with increasing contact energy. The overlap tests have shown that sufficient overlap along a line is necessary for a homogeneous microstructure. This effect can be enhanced by overlapping in both plane directions, whereby the penetration depth of the influenced microstructure also increases from z_{pl} = 850 µm to 1,000 µm. The comparison with real MHP experiments under increased feed rates has shown similar microstructural effects. In particular, the influence of the machining direction under increased feed rates on the microstructure was identified.

In the second part of this chapter, it was shown by means of a validated numerical FE model of the drop tower test rig that the areas underneath the indentations with the highest strains and strain rates are in good agreement with the experimental observed dark areas in the microstructure. Thereby, a smaller hammer head results in higher contact pressure and thus greater strains. The development of the maximum strains for increasing contact energies follows a power law for all hammer head diameters. The maximum strains $\varepsilon_{\mathrm{pl}}$ = 0.1 were detected for the highest contact energies E_{c} = 1,000 mJ. The relatively larger influenced microstructure areas compared to the indentation geometries in case of a single impact was confirmed. Furthermore, it could be shown that the strain rates were between $\dot{\varepsilon}_{\mathrm{pl}}$ = 2,800...13,540 s^{-1} and increased with smaller hammer head diameters. With reference to existing literature [CHEN11B], a mixture of α'-martensite and shear bands is assumed to occur due to the magnitude of the prevailing strain rates, which confirms the experimental results. Once the influence of contact energies of single indentations on the plastomechanical material state were known, **the question arose how a real MHP processing, which is composed of a finite number of interdependent impact results, affects the resulting surface integrity of AISI 304.** The answer to this question is given in Chapter 6.

6 Influence of MHP on surface integrity

Einfluss des MOH auf die Randzonenintegrität

The result of Chapter 5 is an understanding of the strain rates and plastic deformation behavior for single impacts as a function of selected contact energies. Nonetheless, the effect of multiple, overlapping indentations and thus differing contact conditions and energy inputs in real MHP process and their influences on surface integrity remains unanswered. It must be examined whether the findings from the single impact analyses can be transferred to an area-wide MHP processing.

Therefore, in this chapter experimental and numerical investigation for electrodynamic MHP will be conducted. First, in Chapter 6.1, experimental MHP tests are carried out. The surface layer is characterized regarding surface roughness, macro and micro hardness and residual stresses. Furthermore, the microstructure is evaluated using light microscopy. By means of scanning electron microscopy (SEM) and electron backscatter diffraction (EBSD) the grain size and possible phase transformation after MHP are analyzed. Since the dislocation density and further hard to measure material parameters are difficult to obtain experimentally, numerical modelling of the MHP process will be conducted in Chapter 6.2. Thereby, a dislocation-density based model of LEMIALE ET AL. [LEMI10] is implemented in an FE model of electrodynamic MHP using a VUMAT subroutine. A first analysis of the development of dislocation density because of MHP processing is thus made possible. After the extensive experimental and numerical analysis of MHP-treated AISI 304, the results are summarized in Chapter 6.3 and the research question **"How do defined contact energies influence the surface integrity of austenitic stainless steels in the electrodynamic MHP?"** is answered. In Chapter 6.4 a design model was developed based on previous results.

6.1 Experimental analysis of surface layer after robot-based MHP

Experimentelle Analyse der Randzone nach dem roboterbasierten MOH

This section deals with the experimental MHP tests and the associated analysis of the hammered surface layer as a function of the contact energy applied. In addition to the properties of surface integrity roughness, macro- and micro hardness and residual stresses, the analysis of the resulting microstructure is the focus.

6.1.1 MHP test setup and procedure

MOH-Versuchsaufbau und Vorgehensweise

The MHP tests were carried out using the electrodynamic MHP system type 2002 from ACCURAPULS on an industrial robot ABB IRB6660-205/1.9, see Figure 6.1a. AISI 304 specimens with the dimensions 100 x 40 x 30 mm (L x W x H) were used as material, see Figure 6.1b. The specimen were ground plane-parallel on a surface grinding machine BLOHM Planomat HP 620 CNC. Thereby, a ceramic bonded grinding disc, a cutting speed v_{cu} = 30 m·s^{-1} and infeed a_{p} = 0.01 mm to achieve an average roughness

Sa < 1 µm was used. The chemical composition of the specimen is listed in Table 5.1.

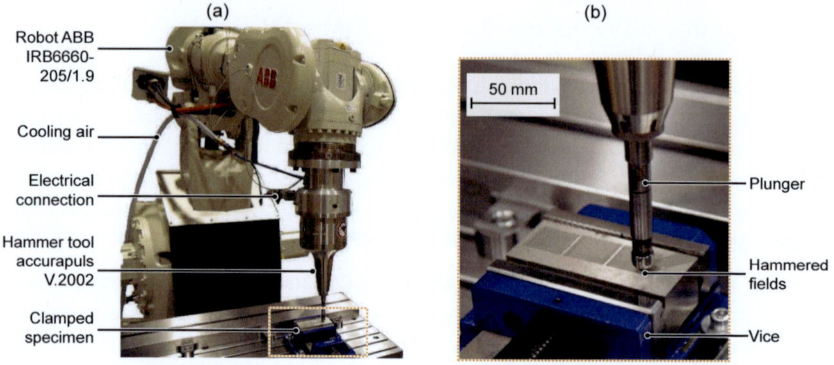

Figure 6.1: Experimental setup of the MHP tests (a) and detailed view of the hammered specimen (b)
Versuchsaufbau der MOH-Versuche (a) und Detailansicht der gehämmerten Proben (b)

The MHP tests were performed using a full factorial experimental design according to the rules of statistical experimental methodology with triple repetition. The hammering frequency f = {120; 180} Hz, the stroke h = {0.3; 0.5} mm, the hammer head diameter d = {6; 12} mm and the indentation distance a = path distance b = {0.05; 0.2} mm were varied in two steps, see Table 6.1.

Table 6.1: Parameter setups for MHP processing of AISI 304 specimens
Parameter-Einstellungen zur MOH-Bearbeitung von AISI 304-Werkstücken

No.	**1**	**2**	**3**	**4**	**5**	**6**	**7**	**8**
a / mm	0.05							
h / mm	0.3				0.5			
d / mm	6		12		6		12	
f / Hz	120	180	120	180	120	180	120	180
E_c / mJ	20	22	20	22	40	32	40	32
w / J·mm^{-2}	8.0	8.8	8.0	8.8	16.0	12.8	16.0	12.8
No.	**9**	**10**	**11**	**12**	**13**	**14**	**15**	**16**
a / mm	0.2							
h / mm	0.3				0.5			
d / mm	6		12		6		12	
f / Hz	120	180	120	180	120	180	120	180
E_c / mJ	20	22	20	22	40	32	40	32
w / J·mm^{-2}	0.50	0.55	0.50	0.55	1.0	0.8	1.0	0.8

All tests were carried out with an impact angle β_i = 0°, which corresponds to an orthogonal alignment of the hammer tool to the specimen surface, and a meander-shaped machining strategy on a field size of 25 x 25 mm. The high-temperature lubricant CASTROL Viscogen KL3 was used as lubricant because of industrial recommendations. One overrun was chosen. The electrical power level was set to P_{el} = 70%. The listed single impact energy E_{c} corresponds to the results from Chapter 4.3. The energy density w was calculated using the following Equation (6.1).

$$w = E_{\mathrm{c}} \cdot \left(\frac{1}{a}\right)^2 \quad \textbf{(6.1)}$$

In order to avoid influences of the hammer head surface on the machining result, it was examined by means of SEM analyses before the MHP tests, see Figure 6.2. Any influence of e.g. imperfections or a flattening of the hammer head surface could be excluded. Only the surface of the hammer head d = 6 mm showed minimal signs of use. The concentric grooves of the hammer head d = 12 mm came from a turning process to produce the plunger. The surface analysis after MHP showed no differences.

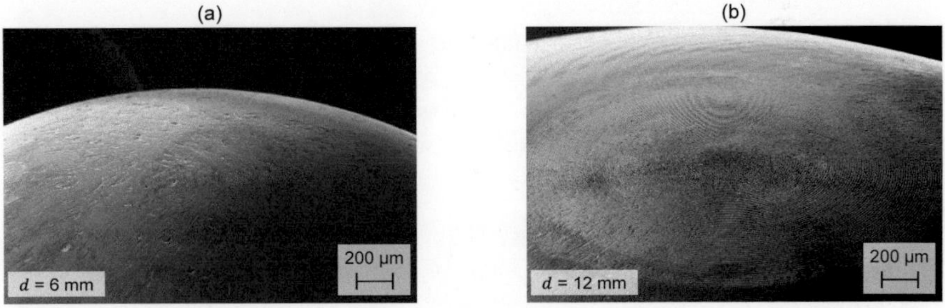

Figure 6.2: SEM analysis of hammer head surfaces d = 6 mm (a) and d = 12 mm (b)
REM-Analyse der Hammerkopfoberflächen d = 6 mm (a) und d = 12 mm (b)

6.1.2 Analysis methods and specimen preparation

Analysemethoden und Probenpräparation

The specimen roughness was determined by means of a measuring system HOMMEL ETAMIC T8000 RC from JENOPTIK AG according to EN ISO 25178 on 1.6 x 1.6 mm measuring fields [DIN12]. The surface macro hardness according to VICKERS was determined by five-fold repeated measurement and subsequent averaging using an INDENTEC ZHU250 hardness tester from ZWICK ROELL according to DIN EN ISO 6507- 1 [DIN18A]. By means of a FISCHERSCOPE HM2000 hardness tester of HELMUT FISCHER GMBH the micro hardness according to VICKERS was measured. The depth profiles of the residual stresses were examined with the PRISM hole drilling system from STRESSTECH GMBH. This measuring device works according to the principle of electronic speckle pattern interferometry (ESPI). The residual stresses are calculated according to the SCHAJER integral method and TIKHONOV regularization [SCHA11].

The AISI 304 microstructure was analyzed by means of light (LM) and scanning electron microscopy (SEM). For microscopic analysis, the specimens were first nickel-plated and then conductively hot-embedded. Grinding was carried out with 320...2500 grit abrasive paper (SiC). Subsequently, the specimens were mechanically polished for 10 minutes each step starting with 6 µm, followed by 3 µm and 1 µm ethanol-based diamond suspension. The selected contact pressure was F_{prep} = 10 N. Colloidal silica suspension (OP-S) was used for the final preparation step to achieve a deformation-free surface for high orientation contrast in EBSD analysis. For LM analysis, the ground and polished specimen cross sections were etched with a BERAHA II solution [BRIT16]. An AXIO IMAGER M2M microscope from CARL ZEISS MICROSCOPY GMBH was used. A SIGMA 500 VP field emission SE microscope from CARL ZEISS MICROSCOPY GMBH was applied for SEM analyses at the CENTRAL FACILITY FOR ELECTRON MICROSCOPY (GfE) of RWTH. Furthermore, for SEM analysis a combined EDX/EBSD system (NORDLYS NANO camera and AZTEC software) from OXFORD INSTRUMENTS was used. The EBSD analysis were performed at an accelerating voltage U_a = 15 kV and a working distance d_w = 18...20 mm.

6.1.3 Results of surface layer characterization

Ergebnisse der Randzonencharakterisierung

The results of the surface layer investigations are presented below. A special focus was set on the microstructure. The analyses allow the derivation of cause-and-effect relationships between MHP process parameters and the resulting surface layer properties of AISI 304. In addition, mechanisms at the microstructure level were identified.

Surface roughness

Figure 6.3a shows the roughness values Sa and Sz as a function of the selected MHP setups. The PARETO effect diagram in Figure 6.3b was created using the MINITAB statistical software.

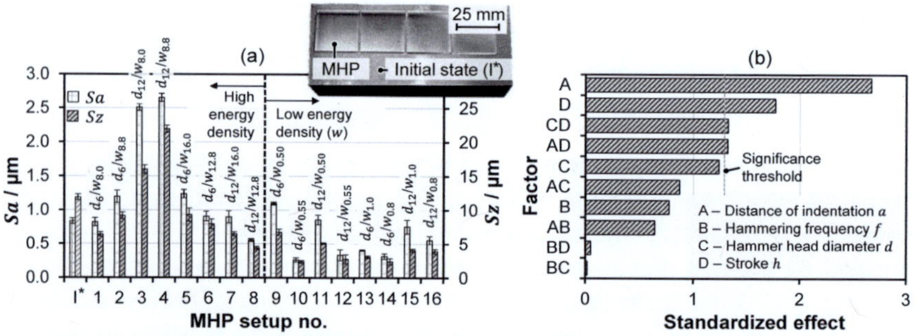

Figure 6.3: Surface roughness Sa and Sz as function of the MHP setups (a) and Pareto chart of effects for Sa (b)

Oberflächenrauheit Sa und Sz in Abhängigkeit von den MOH-Setups (a) und Pareto-Effektdiagramm von Sa (b)

The MHP lead both to an increase and to a reduction of surface roughness in comparison to the ground initial state (I*). However, the already very low roughness of the ground surface Sa = 0.836 µm and Sz = 11.9 µm must be taken into account. The greatest influence has the indentation distance a. Contrary to expectations, an increased indentation distance tends to lead to a reduction of roughness parameters (setup no. 9 - 16). Thus, it is assumed that an indentation distance a = 0.05 mm was selected too small. The increased energy density associated with a smaller indentation a and stepover distance s could therefore be too high for the strength of the AISI 304 stainless steel. This leads to the formation of pronounced surface structures with material build-up. Another explanation could be the higher feed rate associated with a larger indentation distance. This causes an increased flow of the roughness structure in tangential direction, especially for soft to medium hard materials. The strong dependence of the surface roughness on the indentation distance or feed rate was also observed by LIENERT [LIEN17] and BLEICHER ET AL. [BLEI12]. The next largest effect on the roughness values results from the stroke h and the interaction of the hammer head diameter d with the stroke h, see Figure 6.3b. This confirms that the energy density w and the contact pressures are decisive for the processing result. The lowest roughness values Sa = 0.264 µm and Sz = 2.35 µm achieved by MHP setup 10 are even higher than a polished quality. This corresponds to a roughness reduction of 70% and 80% respectively.

Surface layer hardness

The results of surface hardness HV5 are shown in Figure 6.4a. A low indentation force F_i = 49.03 N has been chosen to avoid a martensitic transformation during hardness measurement.

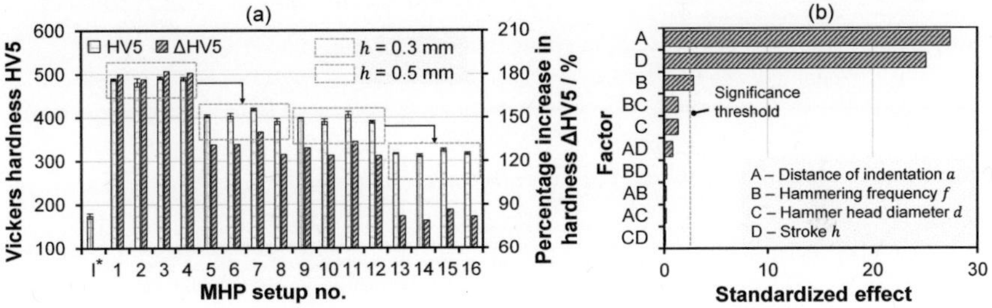

Figure 6.4: Results of macro hardness HV5 as function of the MHP setups (a) and Pareto chart of effects for HV5 (b)

Ergebnisse der Makrohärte HV5 in Abhängigkeit von den MOH-Setups (a) und Pareto-Effektdiagramm von HV5 (b)

For all MHP setups the surface hardness could be increased significantly compared to the initial ground state (I*). For MHP setup 3, a maximum hardness of 491 HV5 could be achieved. This corresponds to a percentage increase in hardness ΔHV5 = 182%. In addition to the indentation distance a, the stroke h and the hammering frequency f also have a significant influence on the surface hardness. A smaller indentation distance a and thus a higher energy density leads to a higher surface hardness. A lower hammering frequency f leads to an increase in surface hardness. Contrary to the work of MANNENS ET AL. on MHP of 100Cr6 (AISI 52100, DIN 1.3505) [MANN20], a higher stroke h and thus a higher contact energy E_c tends to result in a lower surface hardness. These observations coincide with those made by ADJASSOHO ET AL. [ADJA12] and BLEICHER ET AL. [BLEI13] on materials of different initial hardness and heat treatment conditions. They had also determined a threshold value of the stroke, from which the increase in hardness was again less pronounced. One possible explanation for this phenomenon is the exceeding of a critical energy input. Above the threshold value, first annihilation processes of dislocations occur in the surface layer due to a very high dislocation density. This finally reduces the effect of work hardening and thus results in lower hardness increase.

The determined main and interaction effects were transferred into a linear description model according to the methods of design of experiments (DOE). The validation and proof of linearity of this model was done by means of an additional midpoint experiment. The postulated linear relationship between the macro hardness HV5 and the influence variables hammer head diameter d, stroke h, indentation distance a and hammering frequency f can be described by the following Equation (6.2).

$$HV5 = \overline{HV5} + c_1 d + c_2 h + c_3 a + c_4 f + c_5 dh + c_6 da + c_7 df + c_8 ha + c_9 hf + c_{10} af \qquad \textbf{(6.2)}$$

With 401.2 $\overline{HV5}$ the hardness mean value of all measurements.

The procedure for the creation of the description model depends on the work of SIEBERTZ ET AL. [SIEB17]. Due to the two-stage experimental design, only linear dependencies are represented by means of the description model. The model constants in Table 6.2 were determined for the main and interaction effects.

Table 6.2: Parameters of the HV5 description model for main and interaction effects
Parameter des HV5-Beschreibungsmodells für Haupt- und Wechselwirkungseffekte

Main effects				
Parameter	$c_1\ (d)$	$c_2\ (h)$	$c_3\ (a)$	$c_4\ (f)$
Value	2.131	40.560	44.435	4.627

Interaction effects						
Parameter	$c_5\ (d,h)$	$c_6\ (d,a)$	$c_7\ (d,f)$	$c_8\ (h,a)$	$c_9\ (h,f)$	$c_{10}\ (a,f)$
Value	0.181	-0.310	2.131	1.285	-0.381	-0.248

The comparison of the measured and calculated hardness values in Table A.1 in the appendix shows very good agreement. The deviation is $\Delta HV5$ = - 1.92...1.37%. By means of the midpoint experiment (MP) a very good agreement with $\Delta HV5$ = 2.69% could also be found, which confirms the linear correlation.

The results of the micro hardness measurements are presented in Figure 6.5. The first measuring point was set at 50 µm underneath the surface. This avoids falsification of measured values due to displaced material towards the surface. At depth intervals of 75 µm, the other measured values were then recorded. The average values were calculated from eight measurements each.

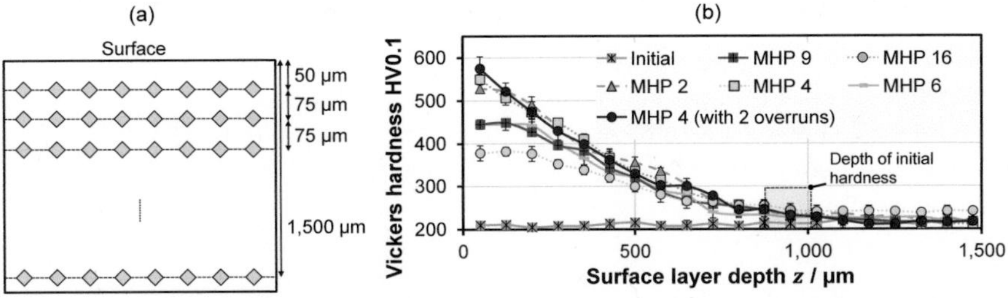

Figure 6.5: Micro hardness measuring grid (a) and results of micro hardness HV0.1 as function of the MHP setups (b)
Mikrohärtemessraster (a) und Ergebnisse der Mikrohärte HV0.1 in Abhängigkeit der MHP-Setups (b)

The micro hardness in the initial state was determined as an average value of 212 ± 5 HV0.1. MHP machining has led to an increase in hardness in the surface layer close to the surface between ΔHV0.1 = 78…171%. The maximum hardness near the surface was 575 HV0.1. One reason for the high increase in hardness is the work hardening of the material brought about by MHP, which in turn is due to a sharp increase in dislocation density. Depending on the energy applied, the formation of subgrain dislocation structures in the form of cell walls and nuclei or arbitrary dislocation arrangements as well as grain refinement can occur. Thereby, the increase in hardness can be explained according to the HALL-PETCH relationship [GOTT07]. Smaller grain sizes lead to an increase in strength. Furthermore, the deformation-induced formation of α'-martensite has an influence on the work hardening behavior and thus an increase in hardness. Because of the small to intermediate strains and strain rates during electrodynamic MHP, it is assumed that α'-martensite resulting from intersections of shear bands consisting of mostly deformation twins is generated. This has also been reported by CHOI and JIN [CHOI97].

Irrespective of the contact energy applied and the contact pressure, the initial microstructure is reached at a surface layer depth z = 800...1,000 µm. The effective depth of the surface layer hardened by MHP is large compared to other mechanical surface layer treatments of AISI 304 [CHER10]. In the range of lower contact energy, the contact pressure has a visible effect on the maximum hardness values. This effect decreases with increasing contact energy. On the other hand, repeated machining of the workpiece surface results in only a slight increase in hardness. This confirms the findings of KRALL for larger hammer head diameters. A reduction of the hammer head diameter tends to lead to a more pronounced influence of multiple machining on the micro hardness [KRAL19]. Furthermore, a combination of high contact energy and high contact pressure does not necessarily lead to maximum work hardening. Dislocation annihilations are likely to occur at this point.

Residual stresses

For the sake of clarity, only two different MHP setups of one of the largest and the lowest compressive residual stress maxima were compared to the ground sample initial state in Figure 6.6a. The characteristic curve of the residual stresses introduced by the MHP was determined for all MHP setups. Nearly all other residual stress graphs of the other MHP setups lie between these curves; see Figure A.6 in the appendix.

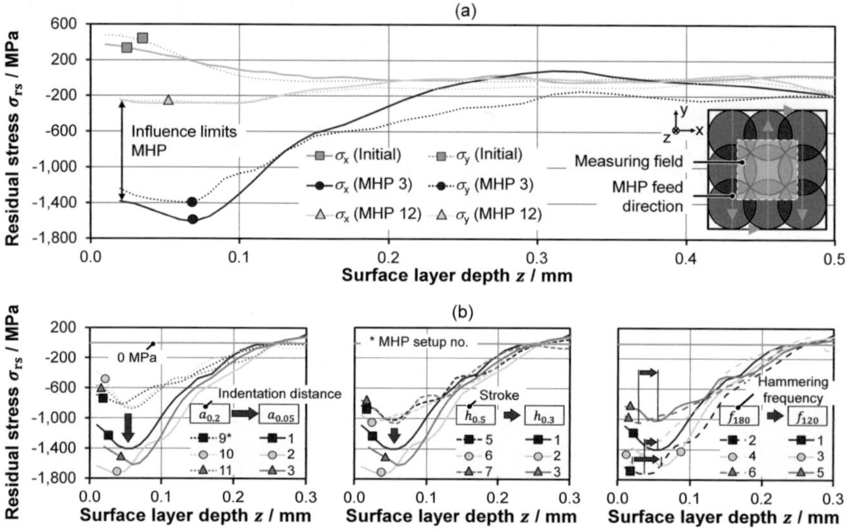

Figure 6.6: Comparison of residual stress distributions after MHP (a) and effect analysis of selected MHP parameters (b)

Vergleich der Eigenspannungsverteilung nach dem MOH (a) und Effektanalyse ausgewählter MOH-Parameter (b)

The residual stresses were measured in the middle of the hammered sample fields (white field, Figure 6.6a). According to the stored coordinate system, the y-coordinate describes the path feed direction of the MHP tool and the x-coordinate the orthogonal direction. As already described several times, e.g. in the work of LECHNER [LECH14], the orthogonal residual stresses are greater than parallel residual stresses. As a result of MHP processing in paths, the material undergoes multiple deformation orthogonal to the feed direction. The compressive residual stresses are correspondingly higher.

For all MHP setups, the maximum of the compressive residual stresses was applied directly below the specimen surface in a surface layer depth z = 0.04...0.1 mm. The HERTZ pressure dominates the multi-axial stress state in the workpiece surface layer resulting from a high normal and lower tangential force component during MHP. In addition, the zero crossing of the residual stresses was already achieved at a surface layer depth z = 0.25...0.35 mm. The effective depth of work hardening and compressive residual stresses therefore differ. The compressive residual stress maxima after

MHP were in a large range $\sigma_{x,max}$ = -330...-1,800 MPa, see Figure 6.6b. The magnitude of the maximum compressive residual stresses is similar to the work of AMANOV ET AL. [AMAN19] and FU ET AL. [FU15]. Due to distinctive α'-martensite transformation as a result of MHP processing and the associated pronounced microstructural stresses over a wider range of the material, significant compressive residual stresses are detected (residual stresses 2nd type).

In addition, there is a strong increase in dislocation density and the associated pronounced stress fields (residual stresses of the 3rd type). The residual stress curves show their maximum at depths comparable to those of deep rolled AISI 304 materials [NIKI07]. A reduction of the indentation distance a = 0.2...0.05 mm and a reduction of the stroke h = 0.5...0.3 mm lead to higher compressive residual stresses. A reduction of the hammering frequency f = 180 to 120 Hz shifts the compressive residual stress maximum to greater surface layer depths. Due to the high energy density resulting from a small indentation distance, beside a partial α'-martensite transformation, a critical threshold value of dislocation density is already reached. The additional increase of the stroke and thus of the impact energy leads to annihilation processes of dislocations and thus to less pronounced compressive residual stresses.

In order to specify the influence of individual MHP process parameters on the maximum compressive residual stresses, effect diagrams were created in Figure 6.7a. An additional PARETO diagram of the standardized effects provides information about statistically significant MHP process parameters; see Figure 6.7b.

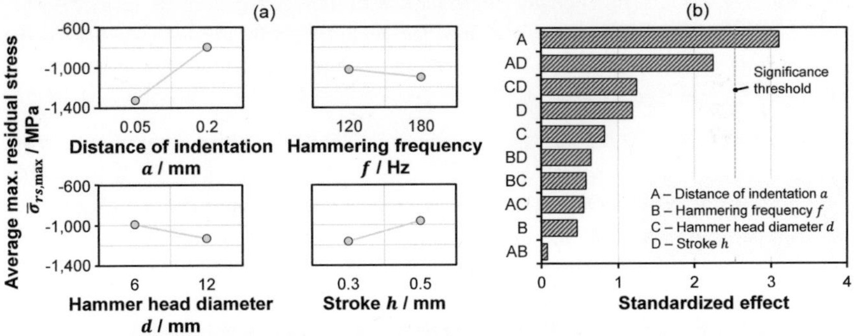

Figure 6.7: Main effect diagrams for maximum residual stresses (a) and PARETO chart of standardized effects

Haupteffektdiagramme der maximalen Eigenspannungen (a) und PARETO-Effektdiagramm der maximalen Eigenspannungen (b)

The indentation distance a and the stroke h had the greatest influence. By reducing them, the maximum residual stresses increase significantly. A larger hammer head diameter d and higher hammering frequency f tend to result in higher stress maxima.

Microstructure (light microscopy)

After measuring the residual stresses, the microstructure resulting from the MHP was analyzed. For this purpose, etched micro sections were prepared using Beraha II, see Figure 6.8. This etching solution makes it possible to visualize the plastically deformed surface layer as well as phase transformations in austenitic stainless steels. Nikitin [NIKI07] and Hotz et al. [HOTZ19] have chosen a similar approach for analyzing microstructural changes in stainless steel.

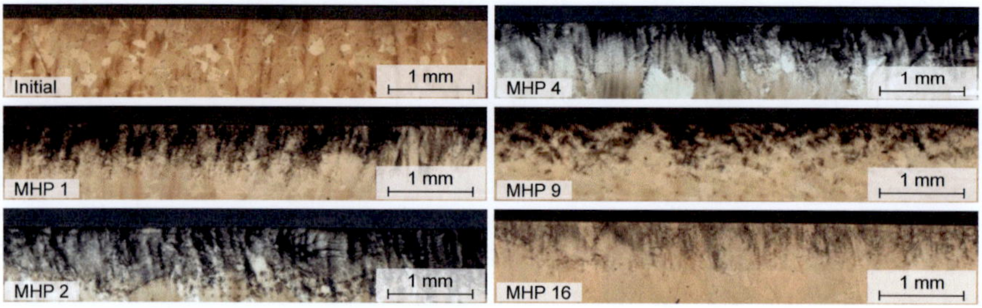

Figure 6.8: Analysis of the plastically deformed surface layer for different MHP setups
Analyse der plastisch deformierten Randzone für unterschiedliche MOH-Setups

Compared to the unhammered initial sample, the influence on the surface layer in the form of dark areas is clearly visible. The dark zone describes the α'-martensite resulting from shear bands and deformation twins. A high dislocation density also characterizes this area. The dislocation activity decreases with increasing surface layer depth. A larger hammer head diameter d = 12 mm (MHP 4) to d = 6 mm (MHP 2) results in a denser transformation zone with the same energy input. However, this zone does not reach very far into the material.

A lower energy input at high surface pressure (MHP 9) creates a very uneven transformation zone. The energy density is therefore not high enough. However, the high surface pressure resulting from the small hammer head diameter d = 6 mm allows the setting of comparable effective depths. Depending on the forming work, the dark area extends to surface layer depths z = 700...850 µm. This nearly corresponds to the measured depth of the hardened layer. MHP setup 16 with a lower energy density and a lower surface pressure (d = 12 mm), for example, shows an only weakly pronounced transformed microstructure. This correlates with the lowest increase in micro hardness. Depending on the forming work and contact pressures, different degrees of forming areas have formed below the surface. No continuous, homogeneous distribution could be achieved. A partial venous formation of areas with high deformation percentages was observed. It needs to be mentioned that using LM, the determination of α'-martensite content is only an approximation since the preparation can also cause a few shear bands and ε-martensite, which lead to a possible overestimation [CELA17].

The influence of different contact energies on the transformation zone is presented in Figure 6.9. A high (MHP 1, w = 8 J/mm²) and a comparatively low energy density (MHP 16, w = 0.5 J/mm²) were analyzed.

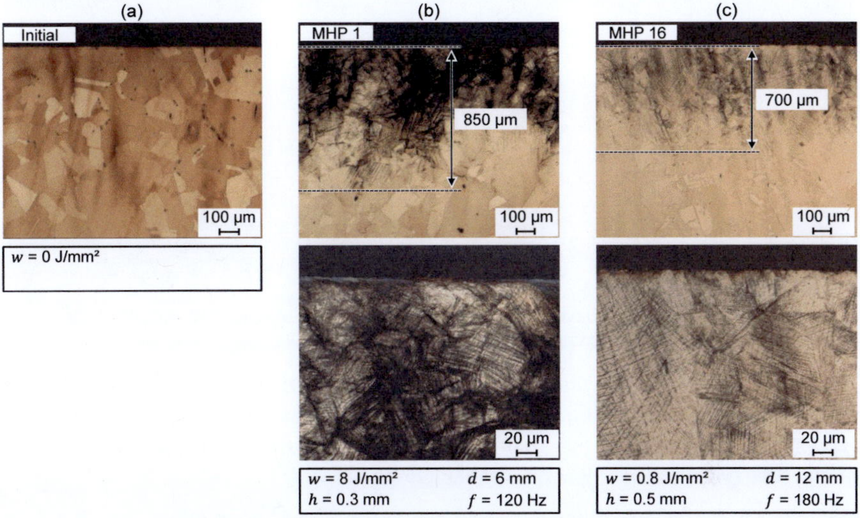

Figure 6.9: Detail comparison of the microstructure in the surface layer of the initial state (a), MHP-setup 1(b) and MHP-setup 16 (c)
Detailvergleich der Mikrostruktur in der Randzone im Ausgangszustand (a), von MOH-Setup 1 (b) und MOH-Setup 16 (c)

Crossing shear bands have formed in both surface layer. In individual grains the formation of micro-twins occurred. A lower energy input greatly reduces the density and homogeneity of the transformed zone. Furthermore, it is found that a higher contact pressure increases the effective depth of the transformation zone. It is assumed that a more uniform transformation zone can be achieved by multiple machining and, above all, a smaller hammer head diameter. The formation of nanocrystals on the hammered surface could not be detected by light microscopic analyses. Furthermore, the proportions of the martensitic phase in the surface layer are still unknown. For this reason, the results of EBSD analyses are presented below.

Microstructure (scanning electron microscopy)

By means of EBSD analyses, conclusions were made about grain size distributions, qualitative dislocation development and phase fractions in the microstructure. Thereby, IPF (Inverse Pole Figure) maps allow for analyzing the orientation and the size of grains. KAM (Kernel Average Misorientation) maps indicate the misorientations between adjacent measuring points and can provide a suitable quantitative estimation of the internal distortions. The total distortions are displayed independently of their dislocations or sub granular origin. Therefore, KAM maps can be used for analyzing the dislocation densities [MOUS15]. Furthermore, phase maps were used for determining a martensite transformation. Figure 6.10 shows the IPF (a), KAM (b) and phase (c) maps for the initial surface layer of AISI 304. A step size of n_{EBSD} = 5 nm was chosen for the following two figures.

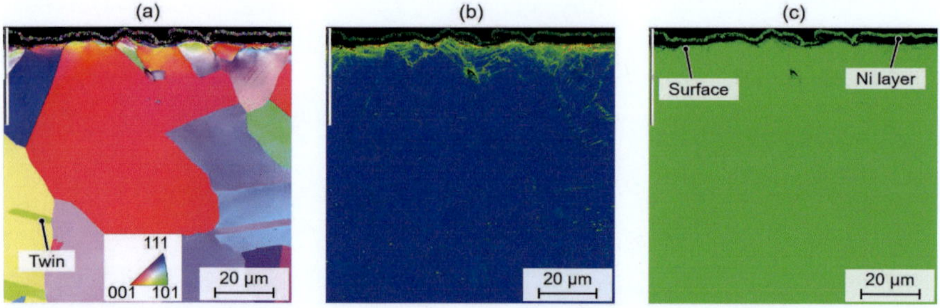

Figure 6.10: IPF (a), KAM (b) and phase (c) maps of the initial AISI 304 surface layer
IPF- (a), KAM- (b) und Phasenbild (c) der AISI 304-Ausgangsrandzone

The colors in Figure 6.10a correspond to the crystallographic orientation in axial MHP impact direction. All following IPF maps show the same orientation. This orientation is represented by means of the color-coded triangle. A few γ-twins appear inside the grains. The grain size distribution is inhomogeneous.

Figure 6.10b shows only at the surface a higher distortion (green areas), which results from the preparation process. In the bulk material no highly distorted regions appear. Thus, the dislocations are homogeneously distributed. Figure 6.10c reveals that the initial microstructure is almost exclusively austenitic (green color). Only at a few places in the microstructure, very small amounts of α'-martensite were detected (red particles). ε-martensite was only perceived in the statistical fluctuation range and could not be determined with certainty.

In order to show the influence of MHP on the AISI 304 microstructure, different surface layer depths of a hammered specimen (MHP 3) has been analyzed. Therefore, Figure 6.11 shows the grain orientations (a) and distortions (b) in different surface layer depth. Significant sub grain refinement of the surface layer to a depth z = 200 µm could be demonstrated. Nanocrystalline grains appear in the region next to the hammered surface. In a surface layer depth z = 200...500 µm, pronounced cubic face-centered twin grain boundaries were generated. At a surface layer depth z = 1,000 µm, a grain distribution similar to the initial unhammered condition was determined.

Based on the KAM maps in Figure 6.11b, it is assumed that there has been an increase in dislocation density in the near-surface layers. The dislocation density decreases with increasing surface layer depth. In the near-surface layers, the dislocations are distributed in a disjointed manner. In a surface layer depth z = 400...500 µm a net-like dislocation arrangement was observed. Crossing shear bands are visible. At a depth $z \approx$ 1,000 µm, a structure similar to the initial state was detected. Decreasing strain rate and strains with increasing surface layer depth appear to correlate with different microstructural features.

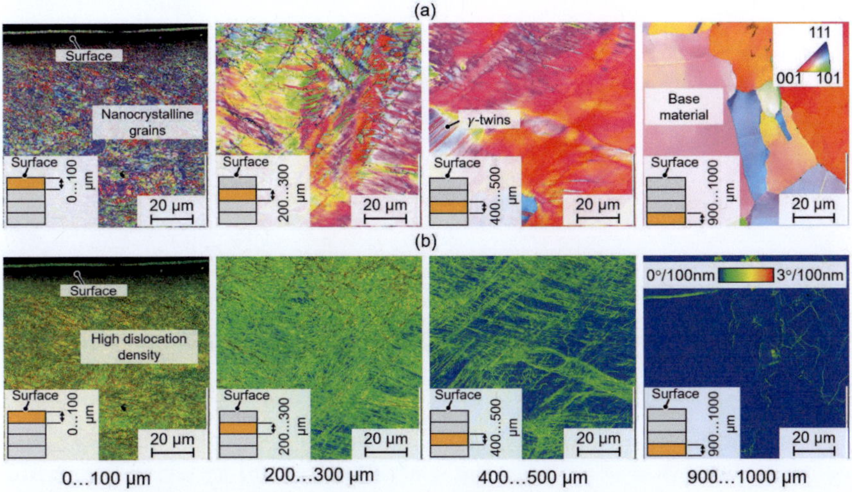

Figure 6.11: Grain size distribution and grain orientation (a) and local misorientations (b) at different surface layer depths after MHP 3
Korngrößenverteilung und Kornorientierung (a) und lokale Fehlorientierungen (b)-Aufnahmen für verschiedene Randzonentiefen nach MOH 3

For analyzing the effect of different contact conditions (energy density, surface pressure) on the surface layer microstructure, the MHP setups 12 and 5 have been compared. These two setups span a wide parameter space used in industrial MHP practice. Furthermore, the number of overruns n = 1; 3 and 5 has been varied. For a better overview, the individual test series are shown in Table 6.3.

Table 6.3: MHP test design for detailed microstructure analysis
MOH-Versuchsplan für detaillierte Mikrostrukturanalyse

No.	5 (1)	5 (3)	5 (5)	12 (1)	12 (3)	12 (5)
a / mm		0.05			0.2	
h / mm		0.5			0.3	
d / mm		6			12	
f / Hz		120			180	
E_c / mJ		40			22	
w / J·mm^{-2}		16			0.55	
n / -	1	3	5	1	3	5

For the analysis of the microstructure after MHP, 100 mm wide and 300 mm deep areas immediately below the hammered surface were analyzed. To reduce the analysis time, a step size n_{EBSD} = 300 nm was chosen. The results of the IPF maps are shown in Figure 6.12.

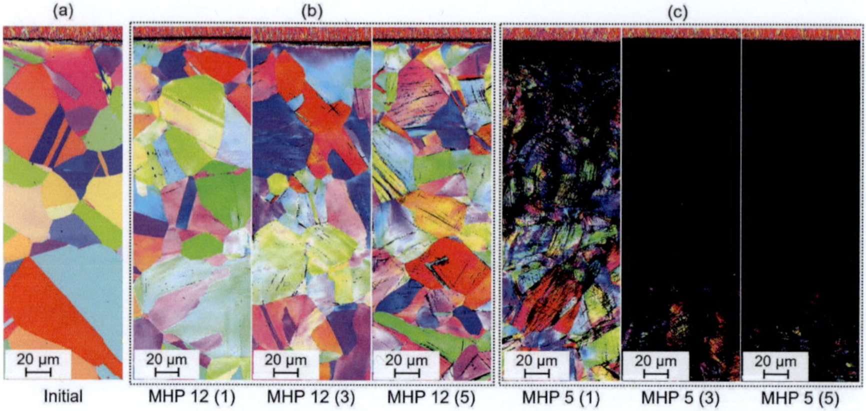

Figure 6.12: Grain size distribution and grain orientation in the near-surface layer for the unhammered initial condition (a), with MHP setup 12 (b) and with MHP setup 5 (c)
Korngrößenverteilung und Kornorientierung in der oberflächennahen Randzone für den ungehämmerten Ausgangszustand (a), mit MOH-Setup 12 (b) und mit MOH-Setup 5 (c)

In the initial state, isolated twin boundaries and few carbide inclusions appeared. Within both test series, dark lines and dark areas are increasingly forming as the number of overruns increases. These were identified as α'-martensite. A low energy input (MHP 12) leads to a significantly weaker martensite transformation. In addition to the lower energy input, the lower contact pressure also contributes to this. Due to a higher energy input (MHP 5), venous-like transformation areas are formed first, as already shown in Figure 6.9. Further overruns lead to an almost complete transformation of the γ-initial structure into α'-martensite. This transformation reaches down to a surface layer depth z = 250 µm. The dark areas indicate a reduction in grain size. Grain sizes were calculated assuming a minimum pixel number of five pixels in the grain and a misorientation of 5° at surrounding grain boundaries. γ-twins were excluded from the grain definition. The grain size distribution for the MHP setups is shown in Figure 6.13.

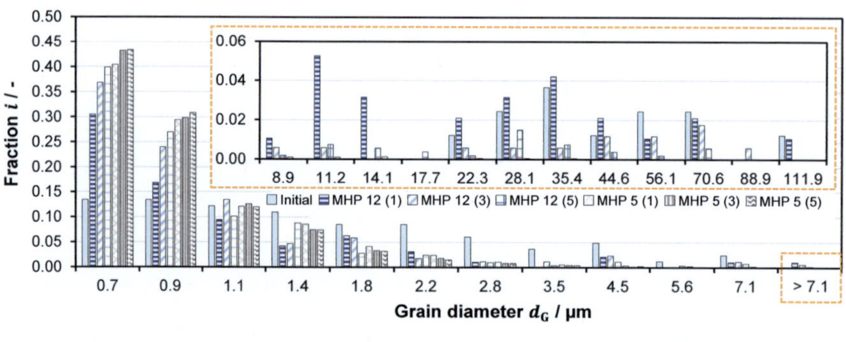

Figure 6.13: Analysis of grain size distribution before and after MHP in the surface layer
Analyse der Korngrößenverteilung vor und nach dem MOH in der Randzone

The specimens hammered with MHP setup 5 show a grain size distribution where most grains are smaller than 5 µm in size. Hereby, the mean grain sizes are in the range $\bar{d}_G$ = 0.48...1.11 µm. In contrast, there are also grain fractions in the range of larger grain sizes for the specimens hammered with MHP setup 12. The average grain size here lies between $\bar{d}_G$ = 2.64...8.90 µm. The average grain size in the initial state was determined to $\bar{d}_G$ = 8.84 µm. However, the results of the grain size distribution should not be over-interpreted. For a statistical verification, more grains would have to be recorded, especially from the coarse-grained initial state. Nevertheless, an at least qualitative proof of the grain size reduction by the MHP can be provided.

Since MHP is a cold forming process in which the recrystallization temperature is not reached, the observed reduction in grain size also describes the formation of subgrains. These result from an increased number of dislocations in different arrangements. For a qualitative analysis of the dislocation development, KAM-maps were generated as described above. These are shown in Figure 6.14.

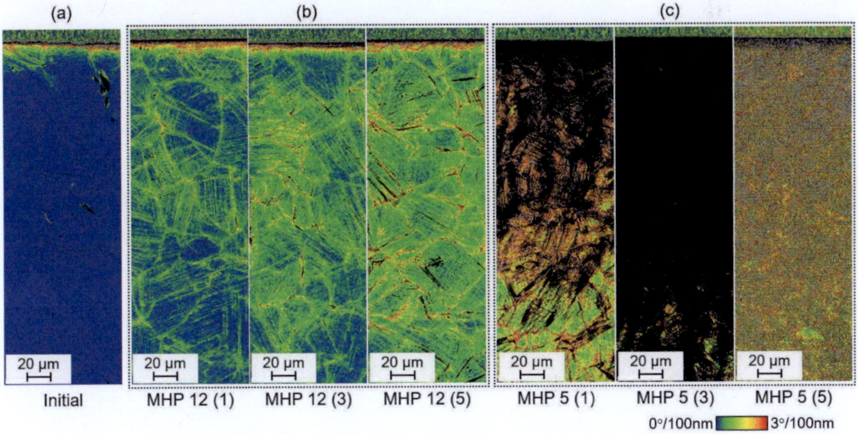

Figure 6.14: Local misorientations in the near-surface layer for the unhammered initial condition (a), with MHP setup 12 (b) and with MHP setup 5 (c)
Lokale Fehorientierung in der oberflächennahen Randzone für den ungehämmerten Ausgangszustand (a), mit MOH-Setup 12 (b) und mit MOH-Setup 5 (c)

The misorientations within an MHP-setup increase with increasing number of overruns. For MHP-setup 12, increased misorientations occur at the grain boundaries. An increased local dislocation density in these areas is the result. This leads to the formation of cell-like structures. A higher number of dislocations in the cell walls than inside the cell characterizes occur. The cell structures are less pronounced than after deep rolling, which is explained by the influence of the forming velocity and strain rate [NIKI07]. The strain rate is higher with MHP than with deep rolling, but lower than with shot peening. In a comparison of the three processes, shot peening therefore has the greatest transverse slip prevention of dislocations. This complicates the formation of spatial dislocation arrangements and leads to mostly stochastic dislocation arrangements.

The misorientation of the specimens hammered with a higher energy density and higher contact pressure is much more pronounced. The dislocation density is again increased. Due to the higher dislocation density, no cell structure is visible. The difference between the KAM maps of MHP 5 (3) and MHP 5 (5) result from a pronounced misorientation between adjacent measuring points. For MHP 5 (5), the change of orientation from one pixel to the next is already so large that there is actually a grain boundary between them. If this happens in all directions, KAM cannot be calculated for such points, because there is no neighboring pixel in the same grain.

Besides the analysis of the grain size distribution and the qualitative development of the dislocation density, the microstructure phases in the surface layer were also examined. The results are shown in Figure 6.15. The γ-austenite appears in green, the α'-martensite in red.

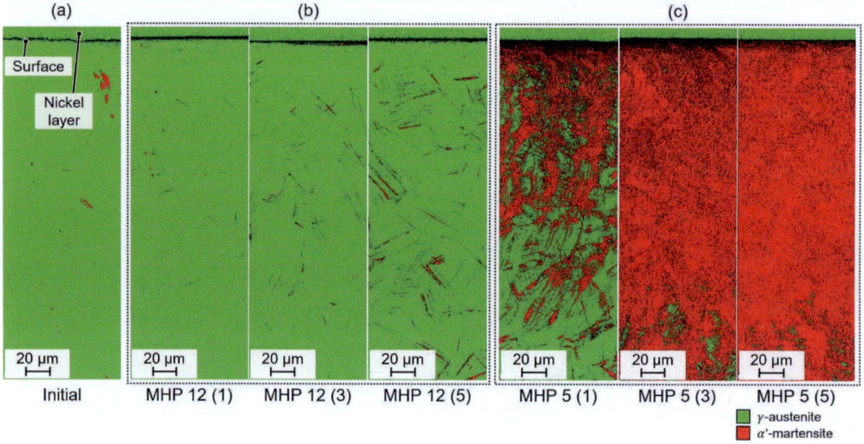

Figure 6.15: Phase determination in the near-surface layer for the unhammered initial condition (a), with MHP setup 12 (b) and with MHP setup 5 (c)
Phasenbestimmung in der oberflächennahen Randzone für den ungehämmerten Ausgangszustand (a), mit MOH-Setup 12 (b) und mit MOH-Setup 5 (c

In comparison to Figure 6.10c it can be seen that at different places in the initial structure also slightly α'-martensite was detected. A low forming energy and comparatively low contact pressure (MHP 12) only leads to a very weak α'-martensite transformation. Mostly parallel lancets characterize the martensite regions. As the number of overruns increases, individual martensite regions are increasingly crossed. In the higher energy range (MHP 5), the venous-like martensite regions already detected in Figure 6.12 form from the surface. With increasing surface layer depth, a previously described structure of crossing martensite lancets occurs more frequently. Several overruns with high forming energy (MHP 5 (3)) lead to almost complete transformation of the surface layer structure into α'-martensite. This area reaches up to a surface layer depth z = 250 µm. A further increase of the number of overruns leads only to a minimal further phase transformation. The fractions of the austenite phase i_γ, the martensite phase i_{α}, the γ-twins i_{Twins} as well as the already mentioned mean grain size $\bar{d}_{\mathrm{G}}$ are listed for the considered specimen states in Table 6.4.

Table 6.4: Overview of phase and twin fractions and average particle sizes
Überblick der Phasen- und Zwillingsanteile sowie mittleren Korngrößen

No.	i_{γ} / -	$i_{\alpha'}$ / -	i_{Twins} / -	$\bar{d}_{\mathrm{G}}$ / µm
Initial	0.995	0.005	0.220	8.841
5 (1)	0.575	0.425	0.033	1.113
5 (3)	0.046	0.954	0	0.488
5 (5)	0.017	0.983	0	0.485
12 (1)	0.998	0.002	0.284	8.904
12 (3)	0.993	0.007	0.082	4.640
12 (5)	0.965	0.035	0.088	2.647

It can be seen that in the initial state of the material there is a purely austenitic microstructure with small amounts of twins and very little martensite. When using MHP setup 5 with a high single impact energy, high energy density and high contact pressure, an increase in the number of overruns leads to a significant increase in the martensite phase content. In addition, a significant grain size reduction is achieved. In the case of MHP setup 12, which is characterized by low energy inputs and contact pressures, multiple processing leads only to a minor α'-martensite transformation. Only a certain influence on the grain size is discernible, but to a much smaller extent than with MHP setup 5. Multiple MHP machining also reduces the proportion of twins in the microstructure. The direct correlation between decreasing twin and increasing martensite fraction suggests that the amount of energy introduced by multiple machining activates different microstructural mechanisms and that a sufficiently high transformation energy density must be provided for martensite transformation.

After the surface integrity of hammered AISI304 workpieces has been extensively analyzed experimentally, there is a need to numerically investigate the underlying stress and strain state. The stress-strain states provide the plastomechanical link between the forming energy introduced and the resulting surface layer properties. Furthermore, the time and cost intensive analyses of the dislocations are to be substituted by qualitative numerical analyses. For this purpose, a numerical FE process model of the MHP is developed in the following, which is coupled with a dislocation density-based material model for analysis of the dislocation density as an internal variable.

6.2 Numerical FE modeling of machine hammer peening

Numerische FE-Modellierung des maschinellen Oberflächenhämmerns

After the experimental analysis of the surface integrity of hammered AISI 304 workpieces, supplementary numerical FE simulations were performed using the ABAQUS 6.14 software. Thereby, the impact velocities and contact energies determined from the distance sensor technology in Chapter 4.3.5 were used for the explicit 3D

MHP model. Thus, for the first time, a realistic energy-based FE analysis of the MHP was made possible. Due to the high expenditure and the large uncertainties in the experimental analysis of dislocation densities, a dislocation density-based material model according to LEMIALE ET AL. [LEMI10] was coupled with the MHP model via a VUMAT subroutine. This material model was validated by means of the experimental flow curves in Chapter 5.3 to ensure a consistent and comparable material behavior in this work. By means of the dislocation density-based material model, the calculation of the dislocation density as an internal numerical variable is enabled. The development of the dislocation density could thus be qualitatively investigated.

6.2.1 Set-up and validation of the FE model

Aufbau und Validierung des FE-Modells

Part modelling. The hammer head was modelled as non-deformable analytical rigid body with an extruded shell. The hammer head diameter d = 6; 12 mm was varied according to the experimental tests. The workpiece was modelled as deformable cuboid with a height h = 5 mm and a square base area with a length l = 6 mm and a width b = 6 mm.

Material properties. The elastic properties of the AISI 304 workpiece were defined with a YOUNG'S modulus $E = 200 \cdot 10^3$ N/mm² and a POISSON'S ratio $\nu = 0.3$. The density of the workpiece was defined as $\rho = 7.9 \cdot 10^{-6}$ kg/mm³. Furthermore, a microstructural constitutive model based on dislocation densities according to LEMIALE ET AL. [LEMI10] was coupled with the MHP FE model using a VUMAT subroutine in ABAQUS software. This material model was developed for impact deformation with high strain rates, which describes similar conditions to MHP. The VUMAT subroutine was written using the programming language FORTRAN 77 and is provided in Figure A.7 in the appendix. The assumptions underlying the material model are described in the following, a detailed explanation is presented by LEMIALE ET AL. [LEMI10] and ESTRIN and KIM [ESTR07].

In the model presented by LEMIALE ET AL., dislocation substructures (cell structures) are formed during material deformation and act as a precursor for a grain refinement. Thereby, the total dislocation density ρ_{tot} is assumed as sum of the two scalar internal variables cell ρ_{c} and wall dislocation density ρ_{w}. A mixing rule is defined, which represents a dislocation migration between dislocation cells and walls. The average dislocation cell size d_{c} and total dislocation density ρ_{tot} is calculated as follow:

$$d_{\mathrm{c}} = \frac{K}{\sqrt{\rho_{\mathrm{tot}}}} \tag{6.3}$$

$$\rho_{\mathrm{tot}} = f_{\mathrm{w}} \cdot \rho_{\mathrm{w}} + (1 - f_{\mathrm{w}}) \cdot \rho_{\mathrm{c}} \tag{6.4}$$

where K is a constant, and f_{w} the volume fraction of dislocation walls. The evolution of f_{w} with different strain states and strain rates is expressed as follow:

$$f_{\mathrm{w}} = f_{\infty} + (f_0 - f_{\infty}) \cdot \exp\left(-\frac{\gamma^{\mathrm{r}}}{\tilde{\gamma}^{\mathrm{r}}}\right) \tag{6.5}$$

with f_0 as the initial value of f_{w} and f_∞ its saturation value. $\tilde{\gamma}^{\mathrm{r}}$ describes the variation of f_{w} with different resolved shear strain γ^{r}. The assumption of $f_0 > f_\infty$ implies that the dislocation walls become sharper as the strain increases in the material.

The evolution of the dislocation densities in the cell walls and interiors is based on dislocation nucleation, interaction and annihilation. To describe these dislocation variations, the following differential equations were developed:

$$\dot{\rho}_{\mathrm{c}} = \alpha^* \frac{1}{\sqrt{3}} \frac{\sqrt{\rho_{\mathrm{w}}}}{b_{\mathrm{B}}} \dot{\gamma}^{\mathrm{r}} - \beta^* \frac{6\dot{\gamma}^{\mathrm{r}}}{b_{\mathrm{B}} d (1 - f_{\mathrm{w}})^{1/3}} - k_0 \left(\frac{\dot{\gamma}^{\mathrm{r}}}{\dot{\gamma}_0}\right)^{-1/n} \dot{\gamma}^{\mathrm{r}} \rho_{\mathrm{c}} \tag{6.6}$$

$$\dot{\rho}_{\mathrm{w}} = \frac{\sqrt{3}\beta^* \dot{\gamma}^{\mathrm{r}} (1 - f_{\mathrm{w}}) \sqrt{\rho_{\mathrm{w}}}}{f_{\mathrm{w}} b_{\mathrm{B}}} + \frac{6\beta^* \dot{\gamma}^{\mathrm{r}} (1 - f_{\mathrm{w}})^{2/3}}{b_{\mathrm{B}} d f_{\mathrm{w}}} - k_0 \left(\frac{\dot{\gamma}^{\mathrm{r}}}{\dot{\gamma}_0}\right)^{-1/n} \dot{\gamma}^{\mathrm{r}} \rho_{\mathrm{w}} \tag{6.7}$$

In the above equations, the assumption was made that the shear rate $\dot{\gamma}^{\mathrm{r}}$ is identical in the cell walls and inside the cells. The first term on the right side of the two equations describes dislocation generation by activating FRANK-READ sources [FRAN50]. The parameters α^*, β^* and k_0 are numerical constants, b_{B} is the size of the Burgers vector. The reduction in dislocation density due to dislocation movement from the cell interior to the cell walls is described by the second term of the equations. The third, negative, term represents dislocation annihilation due to dynamic recovery processes. ρ_{w} and ρ_{c} are the dislocation densities in the cell walls and inside the cell respectively. $\dot{\gamma}_0$ describes the reference shear rate and the exponent n_{L} takes into account the dynamic recovery and is inversely proportional to the absolute temperature [LEMI10]:

$$n_{\mathrm{L}} = \frac{B_{\mathrm{L}}}{T} \tag{6.8}$$

with the constant B, which depends on the stacking fault energy of the material. For FE simulation the mechanical material behavior is detailed with a constitutive model which formulates a relationship between the equivalent (von Mises) stress σ_{vM} and the equivalent plastic strain rate $\dot{\varepsilon}_{\mathrm{pl}}$. The equivalent stress σ_{vM} is written as the sum of two terms:

$$\sigma_{\mathrm{vM}} = \sigma_1 + \sigma_2 \tag{6.9}$$

where σ_1 describes a strain rate-independent component which results from the resistance to dislocation glide. This does not include dislocation-dislocation interactions. σ_2 is the strain rate-dependent component and describes dislocation-dislocation interactions. $\dot{\varepsilon}_{\mathrm{pl}}$ and σ_2 are described by the following equations:

$$\dot{\varepsilon}_{\mathrm{pl}} = \frac{\dot{\gamma}^{\mathrm{r}}}{M} \tag{6.10}$$

$$\sigma_2 = \tau^{\mathrm{r}} \cdot M \tag{6.11}$$

with $\dot{\gamma}^{\mathrm{r}}$ the shear strain rate, M the Taylor factor and τ^{r} the resolved shear stress. Thereby, τ^{r} is obtained by applying a similar rule of mixtures as for the dislocation densities with $\tau_{\mathrm{w}}^{\mathrm{r}}$ and $\tau_{\mathrm{c}}^{\mathrm{r}}$ refer to the resolved shear stresses of the cell walls and cell interiors respectively:

$$\tau^{\mathrm{r}} = f\tau_{\mathrm{w}}^{\mathrm{r}} + (1-f)\tau_{\mathrm{c}}^{\mathrm{r}} \tag{6.12}$$

The parameters $\tau_{\mathrm{w}}^{\mathrm{r}}$ and $\tau_{\mathrm{c}}^{\mathrm{r}}$ are described with the following equations, where α_{L} is a constant, G_{L} is the shear modulus, $\dot{\gamma}_{0,s}$ is a reference shear rate and $1/m_{\mathrm{L}}$ is a strain rate sensitivity parameter:

$$\tau_{\mathrm{w}}^{\mathrm{r}} = \alpha_{\mathrm{L}} G_{\mathrm{L}} b \sqrt{\rho_{\mathrm{w}}} \left(\frac{\dot{\gamma}^{\mathrm{r}}}{\dot{\gamma}_{\mathrm{s},0}}\right)^{1/m_{\mathrm{L}}} \tag{6.13}$$

$$\tau_{\mathrm{c}}^{\mathrm{r}} = \alpha_{\mathrm{L}} G_{\mathrm{L}} b \sqrt{\rho_{\mathrm{c}}} \left(\frac{\dot{\gamma}^{\mathrm{r}}}{\dot{\gamma}_{\mathrm{s},0}}\right)^{1/m_{\mathrm{L}}} \tag{6.14}$$

The temperature dependence of m is inversely proportional to the absolute temperature T, similar to n. Here A is a numerical constant.

$$m = \frac{A}{T} \tag{6.15}$$

Most of the model parameters were obtained from literature sources and are listed in Table 6.5. These parameters can be determined experimentally only at great expense and with the appropriate analysis technology. With the literature values of the individual model parameters used in this work, it was possible to create a starting condition to describe the AISI 304 material behavior. The parameter m was initially set to 120 [DING14] and is continuously overwritten by the calibrated numerical parameters. The parameter $\tilde{\gamma}^{\mathrm{r}}$ = 2.5 s^{-1} was approximated to the values chosen by DING and SHIN [DING12B]. The reference shear rate $\dot{\gamma}_0$ = 100 s^{-1} was set according to the work of WANG ET AL. on shot peening [WANG20A]. The parameter K = 160 was derived from the experimentally determined average grain size $\bar{d}_{\mathrm{G}}$ = 8.8 µm and from the dislocation density in the initial state given in the literature [GRÄN19]. Therefore, in this work, a constant grain/cell size d_{c} is assumed. To determine the parameters α^*, β^* and k_0 of the material model, a MATLAB program was developed The program allows for calculation of the relationship between stresses and strains at an infinitesimal scale based on a uniaxial upsetting test. The calculated flow curves were simulated for one single element at different load conditions (strain rates). The obtained numerical data could thus directly be compared to the experimental data (flow curves). The parameters α^* = 0.22, β^* = 0.01 and k_0 = 19.8 for controlling the dislocation density evolution rate were determined by calibrating the numerical flow curve with the experimental flow curve, see Figure 6.17a. The initial parameters needed to start the iterations were chosen within reasonable limits according to existing literature values [DING12A]. Using the mean square error function, the parameter limits were successively reduced to approximate the numerical to the experimental flow curve. The chosen constants for the dislocation density model are shown in Table 6.5.

Table 6.5: Parameter for dislocation density based material model of AISI 304
Parameter für versetzungsdichten-basiertes Materialmodell von AISI 304

From literature					
Parameter	**Value**	**Source**	**Parameter**	**Value**	**Source**
b / m	$2.56 \cdot 10^{-10}$	[TWAR13]	M / -	3.06	[LI20]
ρ_{w0} / m^{-2}	10^{12}	[GRÄN19]	n / -	100	
ρ_{c0} / m^{-2}	10^{12}		f_0 / -	0.25	
α / -	0.25	[LEMI10]	f_∞ / -	0.06	

Mesh discretization. The workpiece was meshed with four-node tetrahedral element with one integration point (C3D4, $460 \cdot 10^3$ elements) and an element size range l_e = 10…50 µm, see Figure 6.16a. Contact problems can often be realized better with linear than with quadratic elements with only slightly worse accuracy to computational effort ratio.

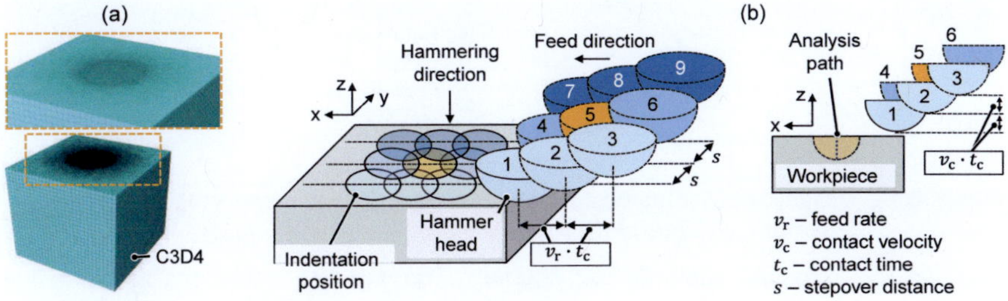

Figure 6.16: FE mesh of the workpiece (a), model structure and analysis methodology (b)
FE-Netz des Werkstücks (a), Modellaufbau und Analysemethodik (b)

The simulations are characterized by strong gradients for stresses and plastic deformations in different regions. Due to the objective to determine the high stress gradients near the contact area as precisely as possible, a high mesh accuracy is needed there. The large element size differences resulting to the unaffected workpiece volume could lead to very strong element distortions when using hexahedral elements. Different element size are practical since the contact areas are fine meshed and the demands on mesh fineness decrease fast with increasing surface layer depth. The use of cuboid elements is critical because of the large distortions. The hammer head was meshed with rigid elements (R3D4/R3D3, $254 \cdot 10^3$ elements) with an element size l_e = 5 µm.

Boundary conditions and process kinematics. All rotational degrees of freedom of the hammer head were set to zero. The translational degree of freedom transverse to the feed direction of the hammer head is also zero. The feed was defined according to the desired indentation distances a = {0.05; 0.2; 0.5} mm. The translational degree of freedom in axial hammering direction corresponds to the impact velocities determined in Chapter 4.3.5. In contrast to TRAUTH's work [TRAU16A], not a single hammer head moves along a path by specifying a sinusoidal motion. In the model presented here, nine hammer heads were defined and positioned to the workpiece surface according to Figure 6.16b. The bottom and outer surfaces of the workpiece were fixed using the *encastre* option. The normal contact condition *surface to surface* was defined between hammer head and workpiece surface. The tangential behavior was defined with a coefficient of friction μ = 0.1, which is suitable for the contact between steel (workpiece) and tungsten carbide (hammer head) [TRAU16A]. The mass of the hammer head was defined as m = 440 g.

Validation. The FE model of MHP with coupled dislocation density-based material model has been verified analyzing the residual stress graphs, see Figure 6.17b. Thereby, the experimental and numerical residual stresses perpendicular to the MHP feed direction were determined.

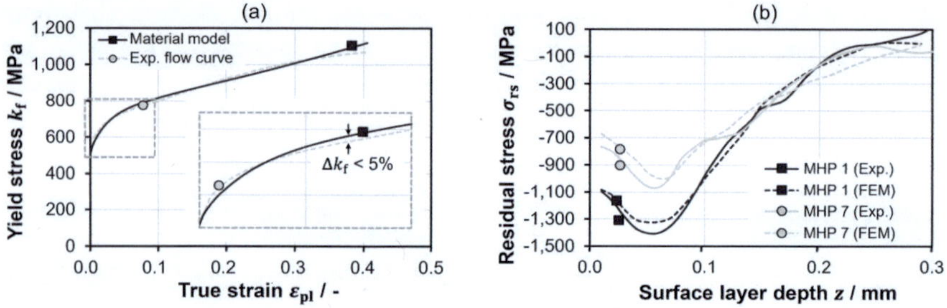

Figure 6.17: Validation of the material model (a) and the MHP FE model (b)
Validierung des Materialmodells (a) und des FE-Modells des MOH (b)

The residual stresses show similar courses. The average deviations between the experimental and numerical values are 5...8%. The lower maximum compressive residual stresses from the simulation are attributed to the missing possibility to map the martensite transformation. The lattice distortions resulting from the martensite transformation and the resulting micro residual stresses were not recorded numerically. Nevertheless, there is a qualitatively satisfactory agreement, which is why the FE model is regarded as validated.

6.2.2 Analysis of stress states and dislocation density

Analyse der Spannungszustände und Versetzungsdichte

Using the calibrated material behavior and the validated numerical FE model of MHP, stress-strain states as well as the resulting dislocation densities were analyzed. Prior to this, however, it was analyzed how the energy fractions behave as a function of selected MHP parameters impact velocity v_c, indentation distance a and hammer head diameter d. The total energy of the system is composed of the frictional energy E_{Fr}, the elastic energy at the workpiece E_{el}, the kinetic energy $E_{kin,S}$ at the beginning of the forming process, the plastic energy on the workpiece E_{pl} and the kinetic energy at the end of the forming process $E_{kin,E}$. These results are shown in Figure 6.18.

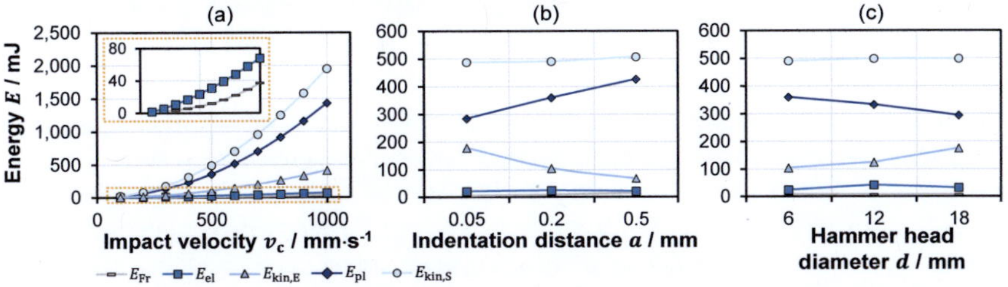

Figure 6.18: Analysis of the MHP energy components as a function of impact velocity (a), indentation distance (b) and hammer head diameter (c)

Analyse der MHP-Energieanteile in Abhängigkeit von der Aufprallgeschwindigkeit (a), des Eindruckabstandes (b) und des Hammerkopfdurchmessers (c)

Figure 6.18a shows that a large part of the kinetic energy is converted into plastic workpiece energy at the beginning of the forming process. In relation, the frictional and elastic energy in the workpiece are negligible. The kinetic energy of the system at the end of the forming process results from the rebound phenomenon of a sphere on a flat surface. In this process, part of the elastic energy is converted again. With increasing impact velocity, the energy fractions follow a polynomial course of second degree. The ratio of kinetic energy to plastic energy is always in the range 1.360...1.365. Figure 6.18b shows that a greater indentation distance lead to a higher plastic energy fraction. This is explained by means of the increasing feed rate for a constant hammering frequency. Thereby, the plastic deformation grows because an increasing tangential force component in the workpiece contact displaces more material in the feed direction. Due to the decreasing contact pressure with increasing hammer head diameter, a reduction of the plastic energy is observed. This is shown in Figure 6.18c. The slight differences in the kinetic energy at the beginning of the forming process result from various ball masses.

Furthermore, the equivalent stresses, the true strain and the dislocation density were analyzed in the following. Figure 6.19 illustrates an example of a hammer head d = 6 mm, ball mass m = 440 g and contact energy E_c = 40 mJ. It shows the equivalent stress σ_{vM} (a), the true strain φ (b) and the total dislocation density ρ_{tot} (c).

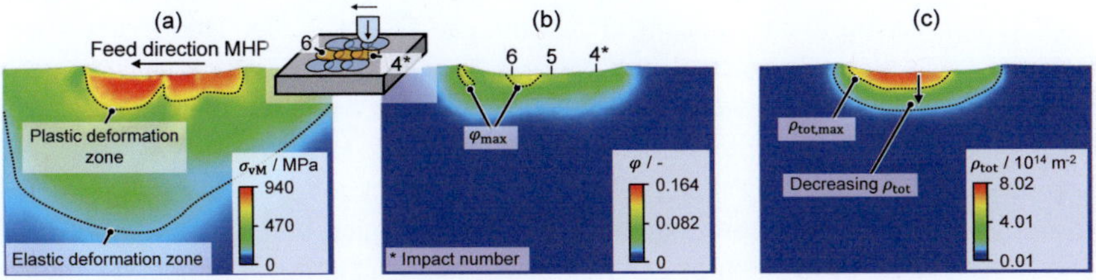

Figure 6.19: Analysis of equivalent stresses (a), true strain (b) and dislocation density (c) after nine MHP impacts

Analyse der Vergleichsspannungen (a), des Umformgrades (b) und der Versetzungsdichte nach neun MOH-Einschlägen

Figure 6.19a shows that in the direction of the MHP feed there is a material stress resulting from the force vector. The maximum equivalent stresses are not homogeneously and symmetrically distributed below the individual indentations. The equivalent stresses increase due to the material throw-up in the feed direction. In comparison to the plastic deformation zone, the elastically influenced material area is large. This also corresponds to the theory according to JOHNSON [JOHN85]. The largest local true strains occur before and below the last indentation of the middle path, see Figure 6.19b. This underlines the results of the equivalent stress and proves the influence of the material raised and displaced in feed direction. The order of magnitude of the true strain achieved by MHP corresponds to the results of TRAUTH [TRAU16A]. The dislocation density in Figure 6.19c shows a similar course. The higher stresses and the pronounced material deformation due to material displacement in the feed direction result in higher dislocation densities. As expected, these dislocation densities decrease with increasing surface layer depth. This is a qualitative result due to the assumptions made about the initial dislocation density.

True strain changes over the surface layer depth are shown in Figure 6.20. The impact velocity v_c, the indentation distance a and the hammer head diameter d were varied.

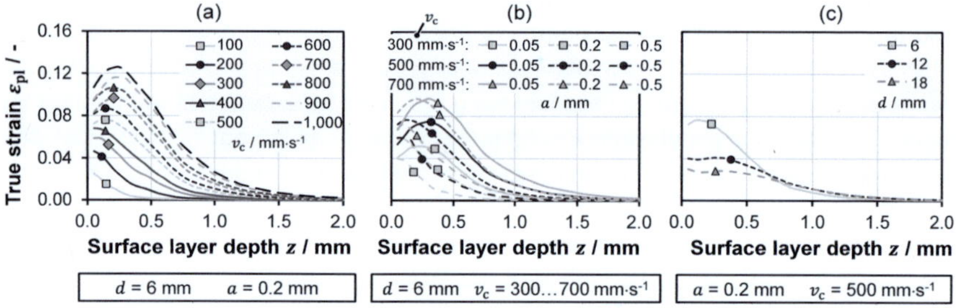

Figure 6.20: True strain as a function of impact velocity (a), indentation distance (b) and hammer head diameter (c)
Entwicklung des Umformgrades in Abhängigkeit der Aufprallgeschwindigkeit (a), des Eindruckabstandes (b) und des Hammerkopfdurchmessers (c)

The maximum true strain shifts from the surface towards greater surface layer depths z < 0.25 mm as the impact velocity increases, see Figure 6.20a. True strain is also increased with higher impact velocity up to φ_{max} = 0.125. The deformations occur in dependence of the impact velocity up to a surface layer depth z = 0.5...2 mm. Figure 6.20b shows that for all impact velocities, the highest true strains are achieved with an indentation distance a = 0.2 mm. A high indentation distance a = 0.5 mm leads to the lowest true strains because of higher material deformation in tangential direction. A decrease in indentation distance generally leads to a shift of the maximum degrees of deformation to greater surface layer depths. Due to the size of the individual indentations of 1 mm, a small indentation distance leads to the fact that a large part of the previous indentation experiences a second energy input. Only a very small proportion of the kinetic energy is transferred to the unprocessed material lying in front of the previous indentation in the form of forming work. Furthermore, a small hammer head diameter leads to a higher pressure. Figure 6.20c shows that the results are larger deformations especially in the direction of the impact axis. The material displacement in tangential direction is higher with a larger hammer head diameter.

The equivalent stresses as a function of the above mentioned influencing variables are shown in Figure 6.21. The individual results can be compared horizontally.

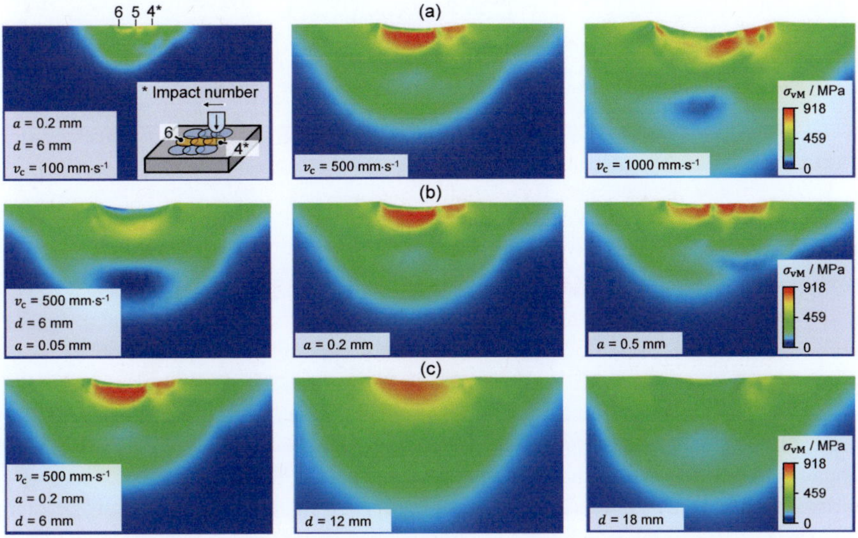

Figure 6.21: Equivalent stresses as a function of impact velocity (a), indentation distance (b) and hammer head diameter (c)

Entwicklung der Vergleichsspannungen in Abhängigkeit von der Aufprallgeschwindigkeit (a), des Eindruckabstandes (b) und des Hammerkopfdurchmessers (c)

With increasing impact velocity, the maximum equivalent stresses increase, see Figure 6.21a. It is noticeable that the maximum equivalent stress is below the first indentation directly at the surface. For the two further indentations, the maximum shifts to a lower-lying surface layer. This effect is intensified with increasing impact velocity. The effective depth of the equivalent stresses increases with increasing impact velocity. The stress distribution is generally very inhomogeneous, which is consistent with the microstructure analyses in Chapter 6.1.3. Figure 6.21b shows that the indentation distance only slightly influences the depth of the elastic deformation zone. A larger indentation distance leads to higher equivalent stresses. The stress maximum shifts towards the surface and spatially separated stress fields are formed. For a homogeneous microstructure influence, a small indentation distance is preferred. Figure 6.21c shows that a larger hammer head diameter leads to lower equivalent stresses. A hammer head diameter d = 12 mm leads to a symmetrical stress field. When using a hammer head diameter d = 18 mm, the maximum stresses occur at the edge of the indentations.

The cause-and-effect relationships between the MHP process parameters and the dislocation density curve were also investigated. The results are shown in Figure 6.22.

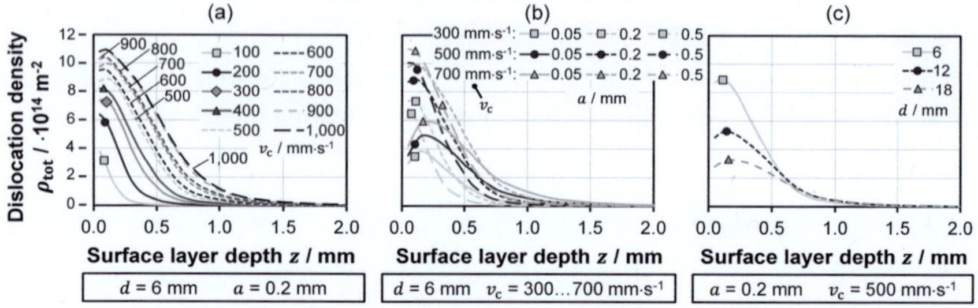

Figure 6.22: Dislocation densities as a function of impact velocity (a), indentation distance (b) and hammer head diameter (c)

Entwicklung der Versetzungsdichten in Abhängigkeit von der Aufprallgeschwindigkeit (a), des Eindruckabstandes (b) und des Hammerkopfdurchmessers (c)

A higher impact velocity leads to greater dislocation densities $\rho_{\text{tot,max}}$ = 11·10^{14} m^{-2}, see Figure 6.22a. With increasing impact velocity, the maximum dislocation density shifts slightly to greater surface layer depths. A corresponding dislocation activity takes place up to a depth z = 0.55...1.5 mm. Figure 6.22b shows that for a given impact velocity, a larger indentation distance leads to higher dislocation densities. This contradicts the findings of the metallographic analyses. One reason could be that only nine impacts were simulated in the FE simulation. In the real MHP, there are considerably more impacts. Due to the large number of impacts, a high, homogeneous dislocation density is only generated with a small indentation distance. The significant influence of a small hammer head diameter on a high dislocation density is shown in Figure 6.22c. However, no significant difference in the effective depth of dislocation activity can be observed. The very high dislocation densities occurring particularly in the surface-near layer indicate the formation of nanocrystalline subgrains. This was also detected in the metallographic images.

6.3 Intermediate conclusion

Zwischenfazit

Chapter 6 dealt with the question **"How do defined contact energies influence the surface integrity of austenitic stainless steels in the electrodynamic MHP?"**

The influence of selected contact energies, energy densities and contact pressures on the surface integrity of AISI 304 was investigated by means of an experimental parameter study on robot-based MHP. Independent of the contact pressure, a low energy density resulted in a significant smoothing of the ground initial surface. Exceeding an energy threshold, which is significantly determined by the stroke and indentation distance, led to surface roughening. The surface hardness was significantly increased by 182% from 175 HV0.5 to 490 HV0.5. A high energy density in combination with a low single impact energy resulted in the highest hardness increases. The hardness increase extended to surface layer depths z = 800...1,000 µm. With the MHP, high compressive residual stress maxima $\sigma_{x,max}$ = -1,800 MPa were demonstrated at surface layer depths z = 0.04...0.1 mm. Higher energy densities as well as lower individual impact energies lead to higher residual compressive stress maxima.

For the same energy input, lower contact pressure resulted in denser martensite transformation in the hammered surface layer. With increasing energy input, the transformation mechanisms of isolated- as well as crossing shear band- and subsequently α'-martensite formation are activated first. The transformation of the surface layer extended to a depth z = 850 µm. EBSD analyses were used to demonstrate that the MHP of AISI 304 leads to subgrain refinement resulting from increased dislocation density. The formation of nanocrystals occurred in the near-surface layer. Depending on the strain rates and strains, various microstructural mechanisms are activated at different surface layer depths. Thereby, the dislocations arrange randomly or in cell-like structures. By means of a numerical FE model of MHP and a coupled dislocation density-based material model, it was confirmed that the dislocation increase in the hammered surface layer. With increasing impact velocity, the maximum dislocation density shifts slightly to greater surface layer depths. To ensure a homogeneous high dislocation density as well as a complete martensite transformation, a sufficiently high single energy input, energy density as well as contact pressure is necessary.

By analyzing the MHP process kinetics, the plastomechanics of individual impacts, and the influence of the MHP on the surface integrity of AISI 304 stainless steel, a comprehensive process understanding was generated. Nonetheless, **the question arose how this knowledge can be synthesized into a design model to increase the fatigue strength of AISI 304 through targeted MHP processing.** The answer to this question is given in Chapter 6.4 and Chapter 7.

6.4 Synthesis of results into design model

Ergebnissynthese zu einem Gestaltungsmodell

The most important results from the chapters 4 to 6 are synthesized in the following in the context of a design model. The knowledge of the cause-and-effect relationships between MHP process parameters and contact energy, between the contact energy and the plastomechanical stress state as well as the understanding of the resulting material-physical mechanisms in the treated surface layer enable a targeted MHP process design.

In **Chapter 4** the MHP kinetics were examined. The results show that the stroke and the electrical power level have a decisive influence on the impact velocity of the plunger on the workpiece surface, see Figure 6.23. An increase of these two parameters rises up the impact velocity and thus the kinetic energy.

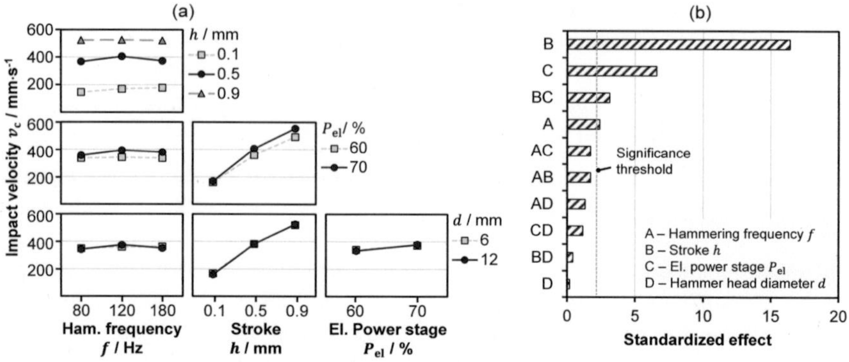

Figure 6.23: Influence of MHP parameters on the impact velocity (a) and derivation of significant influencing variables (b)
Einfluss von MOH-Prozessparametern auf die Aufprallgeschwindigkeit (a) und Ableitung signifikanter Einflussgrößen (b)

Due to the sufficiently high electrical power of the system, small differences in the oscillating mass do not affect the system kinetics. The same applies to a variation of the hammering frequency. Thereby, the impact velocity can be set between v_c = 100...620 mm/s, whereby single impact energies E_c = 3...85 mJ are adjusted. Since the electrodynamic systems differ from each other, the kinetics of each system must be investigated individually. For example, a higher power and a higher oscillating mass will lead to increased kinetic energies. For this reason, in Chapter 5, the resulting indentation geometries as well as the plastomechanical states below the indentations were analyzed for a sufficiently large range of contact energy.

It could be shown in **Chapter 5** that most cause-and-effect relationships between contact energy and geometrical as well as plastomechanical results can be represented by a power function or linear correlations, see Figure 6.24. The approximation functions reproduce the observed behavior well.

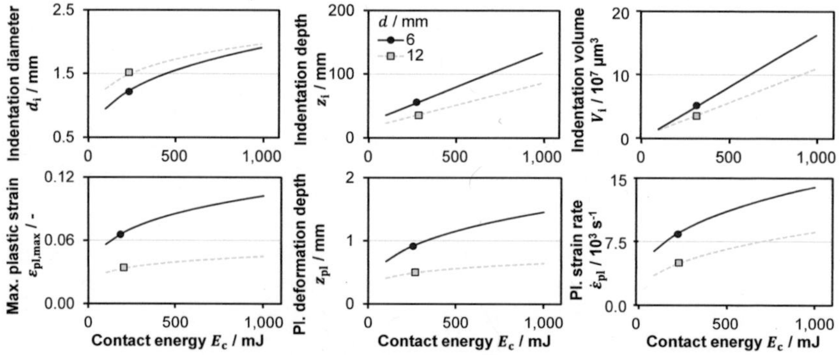

Figure 6.24: Influence of contact energy on geometrical and plastomechanical target values
Einfluss der Kontaktenergie auf geometrische und plastomechanische Zielgrößen

With increasing contact energy, the geometrical indentation dimensions as well as the size of the plastically deformed zone and the strain rates in the workpiece surface layer successively increase. However, the influence of the hammer head diameter has a significant impact on the target values. This is due to the changed contact pressure. While the achievable plastic deformations are very small with $\varepsilon_{\mathrm{pl}}$ = 0.03...0.10, strain rates $\dot{\varepsilon}_{\mathrm{pl}} > 3 \cdot 10^3\ \mathrm{s}^{-1}$ occur. The increase in strain rates resulting from a higher contact energy leads to a more pronounced structural change at greater surface layer depths in the metastable austenitic stainless steel AISI 304 investigated here.

The extensive analyses in **Chapter 6** have shown the influence of the MHP on the processes AISI 304 surface layer. The development of microscopic as well as macroscopic surface layer properties because of MHP processing are shown in Figure 6.25. Figure 6.25a shows the influence of selected MHP parameters on the plastomechanical and microstructural properties of AISI304 stainless steel. True strain, equivalent stress and dislocation densities increase with increasing impact velocity. A small indentation distance also leads to increased dislocation densities. The small contact pressures resulting from a large hammer head diameter lead to less pronounced deformation of the surface layer and thus lower dislocation densities. A higher contact energy of individual impacts as well as an increase in energy density tend to lead to higher work hardening in the surface layer, pronounced martensite transformations and grain refinement. Higher contact pressures lead to an opposite effect.

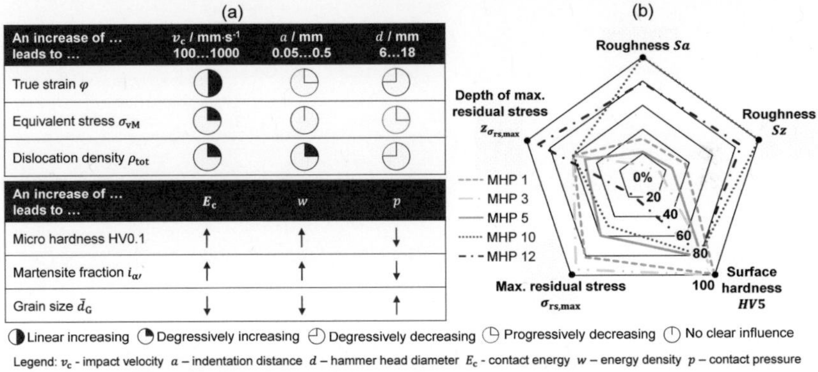

Figure 6.25: Interaction and effect of selected MHP parameters on microscopic (a) and macroscopic (b) surface layer properties
Wechselwirkung und Auswirkung ausgewählter MHP-Parameter auf mikroskopische (a) und makroskopische (b) Randzoneneigenschaften

Figure 6.25b shows exemplarily for five selected MHP setups the scaled influence on macroscopic, easy to measure surface layer properties in the investigated MHP parameter space. The scaling was performed in relation to the highest ($HV5$, $\sigma_{\mathrm{rs,max}}$, $z_{\sigma_{\mathrm{rs,max}}}$) and lowest values ($Sa$, Sz). It can be seen that e.g. MHP setup 10, characterized by a low contact energy, low energy density and high contact pressure, leads to high surface smoothing but low maximum compressive residual stresses. On the other hand, MHP setup 1, characterized by a low contact energy, high energy density and high contact pressure, can be used to set comparatively poor surface roughness but higher compressive residual stress maxima and surface layer hardness.

It is known from the state of the art in cyclic fatigue strength that surface roughness influences crack initiation. A surface as smooth as possible should be aimed for, as this delays crack initiation [KRUP07]. It is also known that a high degree of work hardening in the surface layer also has a positive effect on delaying crack initiation. Therefore, a high hardness of the surface layer is to be aimed for [SCHU06]. Pronounced compressive residual stresses slow down the crack propagation and are less strongly reduced during cyclic loading in combination with high work hardening [TOTT02]. Assuming that the influence of compressive residual stresses and work hardening on fatigue strength predominates in the area of low surface roughness, MHP setup 1 should be used for high fatigue strength. In order to prove that the presented MHP design model helps to select target-oriented parameter for a high fatigue strength, fatigue tests in rotating bending load are performed in the following. By comparing selected MHP setups, the research hypothesis will be validated.

7 Fatigue analysis of MHP-treated surface layers

Ermüdungsanalyse gehämmerter Randzonen

Because of inductive reasoning, in Chapter 4 to 6 the multiphysical system behavior of the electrodynamic MHP was analyzed as well as the contact conditions and surface integrities resulting from varying contact conditions at single or multiple impacts. The underlying physical mechanisms of action and cause-effect relationships between introduced contact energy and surface layer properties were analyzed based on experimental and numerical investigations. The results of Chapter 6 showed that a defined surface integrity of austenitic stainless steel can be set by MHP depending on the applied contact energy. The design model for MHP of austenitic stainless steels derived from the findings of the previous chapters represented a deductive argumentation.

The deductive findings on the targeted, energy-based adjustment of surface integrity in electrodynamic MHP must now be validated. In this chapter, the validation is carried out by means of fatigue investigations. In Chapter 7.1, the MHP test stand is extended by means of an additional rotary axis for manufacturing of rotationally symmetric specimens. These specimens are hammered using different MHP setups, characterized afterwards and the rotating bending fatigue test stand is validated. Chapter 7.2 discusses the results of fatigue strength and fractographs of the broken specimen. Chapter 7.3 summarizes the findings and answers the research question **"What is the relationship between a knowledge-based setting of the surface integrity in the electrodynamic MHP and the resulting fatigue strength?"**

7.1 Experimental design and procedure

Versuchsplanung und -durchführung

After the influences of different contact energies of MHP on the surface layer properties were determined in the previous chapter, it is now necessary to analyze their influence on the application properties of the workpieces. Thereby, material fatigue under constant, cycling loads has a decisive influence on component service life. For this reason, the fatigue strength of hammered AISI 304 surface layer has been investigated experimentally by means of a rotating bending test rig. Since rotationally symmetrical workpieces were tested, the robot-guided MHP test bench was extended by an additional external rotary axis. Subsequently, the MHP tests were carried out using selected MHP system settings from Chapter 6. After MHP treatment, the surface layer were characterized. The hammered workpieces were examined in the rotating bending test bench regarding the number of cycles to failure and breakage pattern. Finally, a recommendation for action was developed.

7.1.1 Extension of MHP test bench and test conditions

Erweiterung des MOH-Prüfstandes und Versuchsbedingungen

Since the fatigue tests were to be carried out on a bending test rig on rotationally symmetrical specimens, the MHP test bench had to be extended to handle the specimen geometry. Therefore, an additional seventh, external rotary axis MTD250 from ABB was installed, see Figure 7.1a.

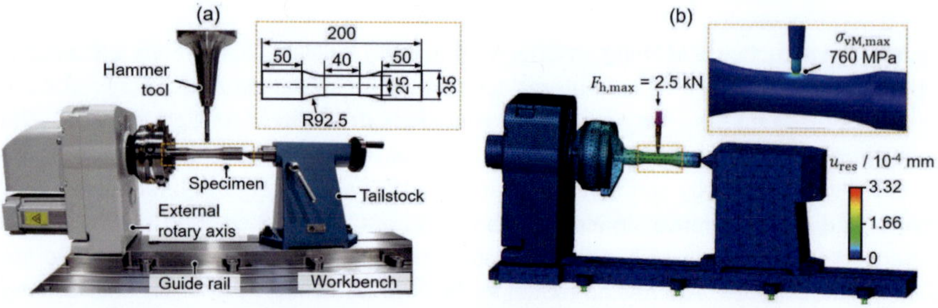

Figure 7.1: MHP test bench extension by means of external rotary axis (a) and FE analysis of the bending load during MHP processing (b)
MOH-Prüfstandserweiterung mittels externer Drehachse (a) und FE-Analyse der Biegebelastung während der MOH-Bearbeitung (b)

A tailstock was used for support on both sides via a centering hole in the front face of the specimens. The tailstock can be specifically positioned for different specimen lengths by means of a guide rail made of 42CrMo4 quenched and tempered steel. The selection and production of the specimen geometry was carried out according to the recommendations of DIN50113 [DIN18B] as well as the specified conditions of the rotating bending test rig. By means of the FE solver of the software SOLIDWORKS, the bending load of the test stand extension was also examined in advance, see Figure 7.1b. Assuming a point load $F_{h,max}$ = 2.5 kN, a maximum deflection of the specimen u_{res} < 1 µm was determined. Thus, a MHP processing of the selected geometry in the extended test bench could be realized without any problems.

The MHP experiments were carried out based on the previous findings on AISI 304 surface integrity, summarized in the design model from Chapter 6.4. To enable a comparison between differently hammered surface layers, three MHP setups were selected, see Table 7.1. These were compared to an unhammered specimen batch.

Table 7.1: MHP test setup for rotating bending specimen
MOH-Versuchsplan für Biegeumlaufproben

MHP setup no.	f / Hz	d / mm	$a = s$ / mm	h / mm	w / J·mm²
1	120	6	0.05	0.3	8.00
10	180	6	0.2	0.3	0.55
16	180	12	0.2	0.5	0.80

MHP setup 10 and 16 has been chosen because of the low surface roughness achieved and because they represent industrial parameter settings close to the energy conditions of MHP setup 12 from chapter 6. All tests were performed with a constant electrical power stage P_{el} = 70%. The MHP was performed in orthogonal position of the hammer tool to the specimen. The lubricant CASTROL Viscogen KL3 was applied for all MHP tests.

The unhammered and hammered surface layers were analyzed before performing the rotating bending tests. The characterization of roughness, surface hardness, residual stresses and hardness course into surface layer depth was performed analogous to Chapter 6.1.2. The results are shown in Figure 7.2.

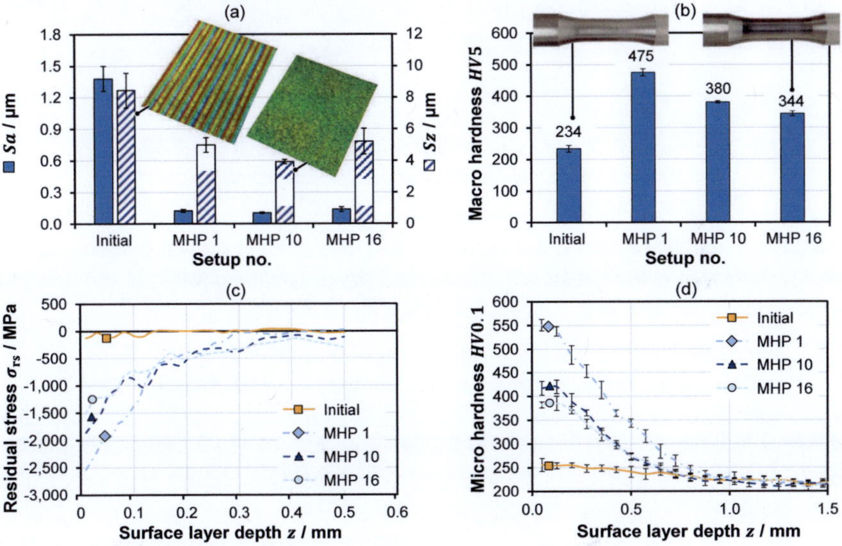

Figure 7.2: Results of surface roughness (a), macro hardness (b), residual stresses (c) and micro hardness (d)

Ergebnisse der Rauheit (a), der Makrohärte (b), der Eigenspannungen (c) und der Mikrohärte (d)

Despite the changed specimen geometry, roughness was achieved comparable to the flat specimens, see Figure 7.2a. With roughness values Sa = 0.126 µm and Sz = 3.93 µm, setup 10 also achieves the lowest roughness. This corresponds to a roughness reduction of 91% and 54% respectively. Similarities to MHP of flat specimen were also shown in the area of surface hardness, see Figure 7.2a. Hardness increases between $\Delta HV5$ = 47...102% could be achieved. Due to the higher compression in contact compared to the MHP of flat specimens, higher compressive residual stress maxima were set. Nevertheless, the absolute values, especially at the surface, should be considered with caution. The analysis of micro hardness confirms the strain hardening pronounced from increased compression.

Furthermore, the microstructure of the fatigue specimens is shown in Figure 7.3. These pictures were taken using the same procedure as in Chapter 6.1.2.

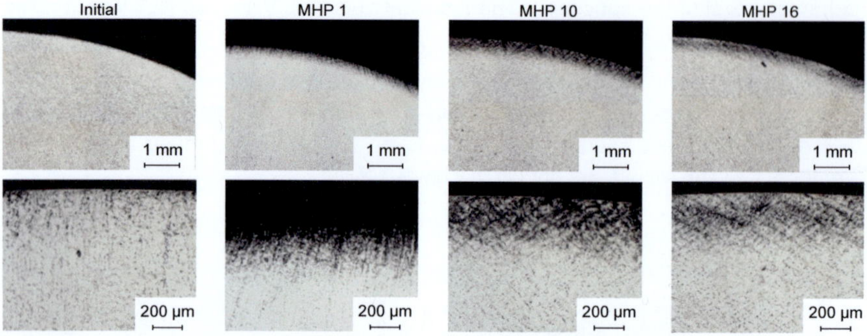

Figure 7.3: Comparison of surface layer microstructure of fatigue specimens
Vergleich der Randzonenmikrostruktur der Ermüdungsproben

MHP setup 1 has resulted in a continuous, transformation surface layer. The other two setups only partially affected the surface layer. The already proven influence of the energy density in combination with the contact pressure during MHP on the resulting microstructure in Chapter 6.1.3 was confirmed again here.

7.1.2 Validation of rotating bending test rig and approach

Validierung des Biegeumlaufprüfstandes und Versuchsdurchführung

The fatigue tests were carried out on a rotating bending test rig from ECOROLL, see Figure 7.4. The test stand was designed according to DIN50113 guideline [DIN18B].

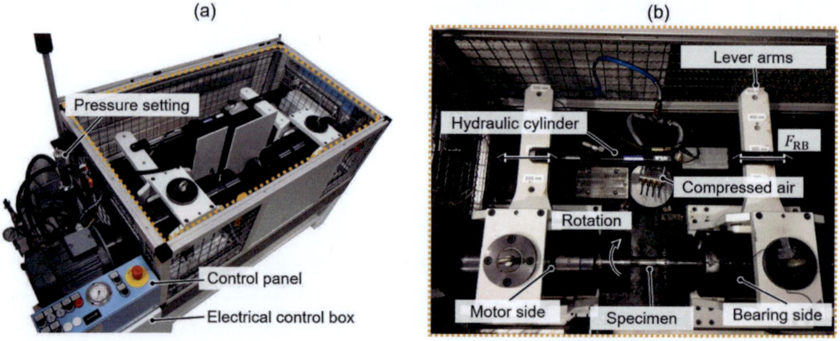

Figure 7.4: Overview (a) and detail view (b) of the rotating bending test rig
Übersicht (a) und Detailansicht (b) des Umlaufbiegeprüfstandes

The specimen are clamped between a drive unit on the motor side and a bearing unit on the guide side. The motor side rotates the specimen. During rotation, a defined bending moment is applied by a hydraulic cylinder, which is connected to adjustable lever arms. The lever arms can be varied in 100 mm steps between 200...500 mm. The required test force is set via the variable pressure on a compact pump unit. A maximum bending moment M_b = 7,500 Nm is therefore adjustable. The specimens can be cooled by using compressed air.

The test bench was calibrated before the experiments. The calibration procedure has been modified compared to the recommendation in DIN50113. The strain gauge attached to a calibration specimen was calibrated by means of a force sensor in a lathe DMG MORI NEF 600, see Figure 7.5a.

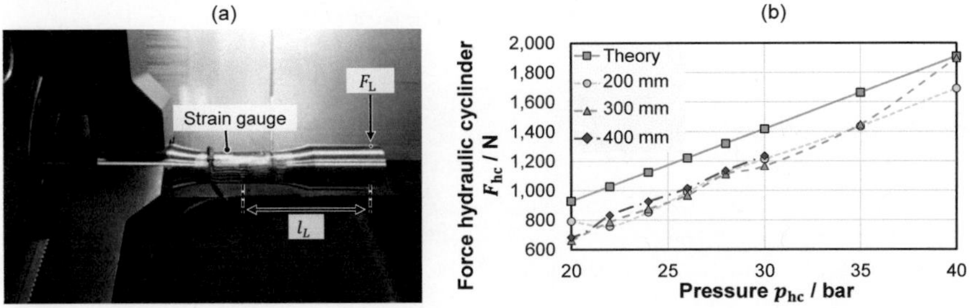

Figure 7.5: Calibration of strain gages in lathe (a) and determined forces in rotating bending test rig (b)
Kalibrierung der DMS in Drehmaschine (a) und ermittelte Kräfte im Biegeumlaufprüfstand (b)

By means of the applied, known force of the lathe F_{Lathe} and the lever arm l_{L}, the resulting bending moment at the strain gauge could be calculated. By knowing the lever arm in the rotating bending test bench, it was possible to calculate back from the calibrated strain gauges to the applied force. Since the test rig showed a behavior that deviated from the theory, see Figure 7.5b, the calibrated strain gauges were used to determine compensation curves that allowed the correct settings for pressure and lever arm to be selected for a defined bending stress. The correction factors were realized using a nonlinear regression.

The fatigue tests were performed according to DIN 50100 [DIN16]. Between specimen and clamping device, additional centering seats BIKON 2006-035-065 were used for concentric and bending moment free specimen fixation. The test frequency was constant at f_{RB} = 25 s^{-1}. A sinusoidal alternating stress with a loading ratio R = -1 was chosen. N_{L} = 10^{6} load cycles were defined as the number of cycles to failure in the fatigue strength range. This value was chosen according to the work of MUÑOZ-CUBILLOS ET AL. [MUÑO17]. The test duration when reaching the defined number of cycles to failure was about 11 hours.

Due to a limited number of specimen, the fatigue strength was determined using a modified staircase testing method by MÜLLER ET AL. [MÜLL12]. This method is based on the findings of RANSOM and MEHL [RANS53]. The novel method allows the reuse of defect-free specimens to increase the accuracy of fatigue strength estimation for small specimen sizes. In equidistant steps (stairs), the bending stress of the following test was adjusted in dependence of a break (failure before N_{L}) or pass through (no failure before N_{L}). The bending stress was increased or reduced in steps of $\Delta\sigma_{\mathrm{B}}$ = 18.5 MPa. Based on the results of the work of NIKITIN [NIKI07], MUÑOZ-CUBILLOS ET AL. [MUÑO17] and SINGH ET AL. [SING17], an initial bending stress $\sigma_{B,0}$ = 240 MPa was selected for the unhammered AISI304 rotating bending specimens.

In preliminary tests without cooling, the specimen break already occurred between N_L = 5...32·10^3 load cycles. The samples showed a clear external discoloration, see Figure 7.6a. This indicated a strong heating of the specimen.

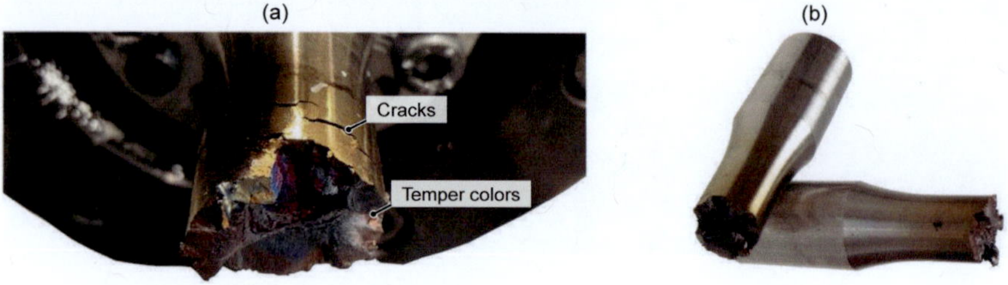

Figure 7.6: Annealing colors in fracture surface (a) and fracture parts (b)
Anlauffarben in Bruchfläche (a) und Bruchhälften (b)

The dark purple color in the fracture surface implies a short-term maximum temperature T > 400 °C. KONGTHEP reports similar results [KONG19]. Figure 7.6b shows that the break occurred in the middle of the specimen. One reason for the specimen heating is the test frequency specified by the test stand. To improve heat dissipation during the tests, the specimens were cooled by compressed air. Random temperature measurements showed that the surface temperatures did not exceed T = 50 °C due to cooling.

7.2 Analysis of load cycles and fractographs

Analyse der Lastspielzahlen und Bruchflächen

The results of the rotating bending tests are presented below. For the analysis of the critical bending stress before failure $\sigma_{b,L}$ of the different specimen states, the evaluation according to MÜLLER ET AL. [MÜLL12], based on the work of HÜCK [HÜCK83] was used:

$$\sigma_{b,L} = \sigma_{b,0} \cdot d_L^{\left(\frac{A_L}{F}\right)} \tag{7.1}$$

with $\sigma_{b,0}$ the lowest bending stresses from evaluable tests and d the logarithmic step change. This was determined by forming the mean value of all step increments to d = 1.067. The factor F describes the sum of the events f_i on all evaluable load steps. The factor A_L is the sum of the products of the number of results on one stress level with the corresponding number of the stress level i. The lowest valuable level receives the index i = 0. For clarification, both factors are described with the following equations:

$$F = \sum f_i \tag{7.2}$$

$$A_L = \sum i \cdot f_i \tag{7.3}$$

The results of the staircase fatigue analysis are shown in Figure 7.7. Pass-through specimens were used a maximum of two more times. At the end of each test series, another fictional test on the coming stress level was counted. Since the applied methodology does not differentiate between a break and without breakage, further information could be included in the evaluation. A similar approach has already been used by TRAUTH ET AL. [TRAU16B].

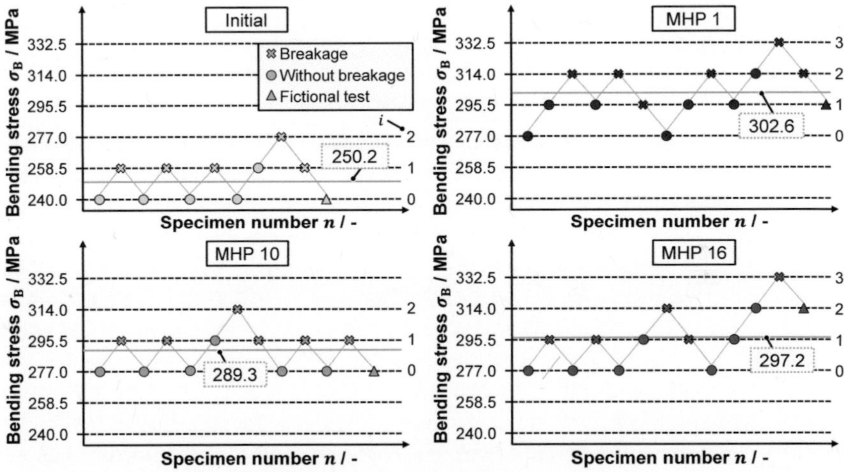

Figure 7.7: Results of the 50% fatigue strength by staircase test
Ergebnisse der 50%-Dauerfestigkeiten mittels Treppenstufenverfahren

The critical bending stress of the unhammered samples was mostly between the two load levels σ_b = 240...258.5 MPa. The 50%-fatigue strength is calculated to $\sigma_{b,crit}$ = 250.2 MPa. The specimen breaks occurred between N = 48...257·10^3 load cycles. All three MHP setups investigated led to an increase in the critical bending stress. Individual preliminary tests showed that a first stress level $\sigma_{b,0}$ = 277 MPa is chosen reasonably. The MHP setup 1 showed the greatest influence on the increase of the critical bending stress $\sigma_{b,crit}$ = 302.6 MPa. This corresponds to an increase by a factor of 1.21.

Using the other two MHP-Setups 10 and 16, increases by a factor of 1.15 and 1.18 were achieved. Therefore, it is proven, that MHP has a positive influence on the fatigue strength of AISI 304 under cyclic loadings. The results suggest that under the conditions considered, the influence of work hardening and compressive residual stresses overlap the influence of surface roughness. Nevertheless, it is believed that an even higher increase of the fatigue strength of the MHP can be achieved by several MHP overruns. However, the resulting higher time expenditure must be taken into account.

Figure 7.8 shows the fracture surfaces (fractographs) of the rotating bending samples. The images were taken with a scanning electron microscope Sigma 500 VP from ZEISS in secondary electron mode.

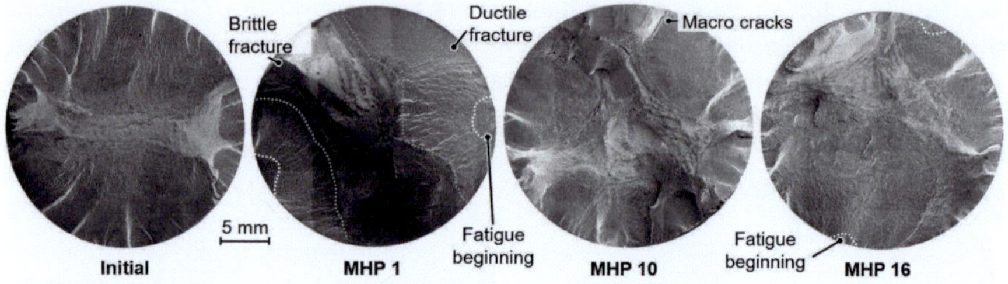

Figure 7.8: Fractographs of different specimen processing conditions
Bruchbilder unterschiedlicher Probenbearbeitungszustände

The fractographs of the initial specimen state showed that fatigue fractures were initiated at the specimen surface. A few micro cracks then propagated into the specimen interior. Fracture paths and a partly fibrous structure is recognizable. In the central specimen area, a bright broad stripe has formed over almost the entire diameter. This corresponds to a brittle residual fracture. The specimen hammered with the highest energy (MHP 1) showed different areas in the fractograph. In the encircled white areas, lines of detent were detected. This was correlated with a multiple crack arrest at the beginning and thus a fatigue fracture originating from these spots at the surface. Areas with striation marks and with a honeycomb like structure occurred. This indicates a hybrid of transcrystalline sliding fracture and brittle force fracture. Before the fatigue fractures converge in the center of the specimen, an area of brittle fracture has occurred. Similar observations were made for the other two hammered sample conditions. However, the proportion of ductile fracture was higher for MHP 10 and MHP 16. Since significantly more energy was introduced into the workpiece with MHP setup 1, an increased content of transformation martensite is assumed. This is mainly found in the surface layer. Together with a significantly increased dislocation density, there is a pronounced resistance to crack growth. However, when a critical crack expansion energy is exceeded, pronounced brittle fracture occurs due to the more brittle phase properties.

Figure 7.9 shows detailed pictures of the unhammered specimen. The areas under consideration are outlined in color and numbered. In the detailed image of the unhammered specimen, oscillating stripes in different fracture trajectories are visible. These are due to the alternating load. The concave shape of the oscillating stripes coming from the surface underlines the point of origin at the surface. On a microscopic level, cleavage surfaces have appeared. Steps and flow lines have formed. It is assumed that many small incipient cracks have united, resulting in a transcrystalline brittle fracture in the specimen center. The shear lips at the crack end indicate an altered stress state and a macroscopic brittle force fracture in the specimen center. The light veins indicate plastic deformation.

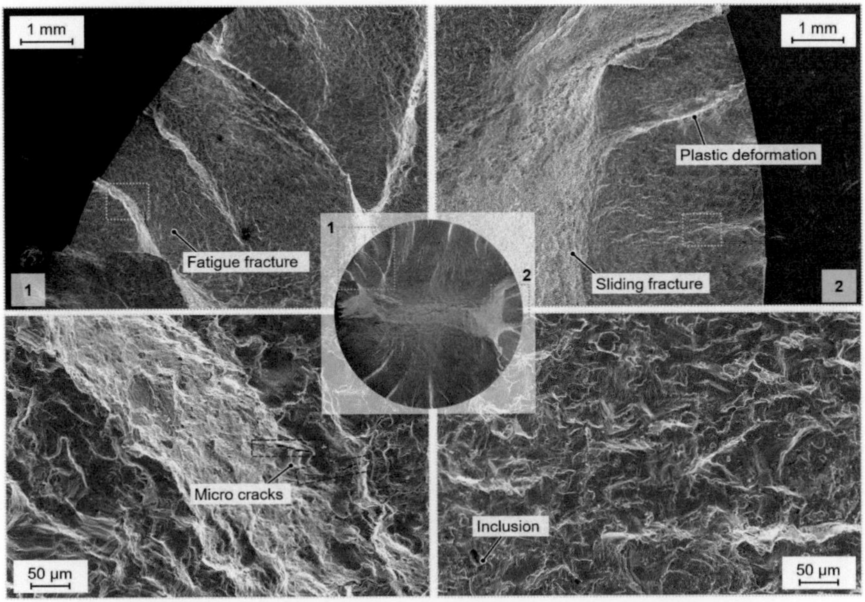

Figure 7.9: Detailed fractographs of untreated AISI 304 specimen (initial)
Detailaufnahme der Bruchbilder von AISI 304-Proben im Ausgangszustand

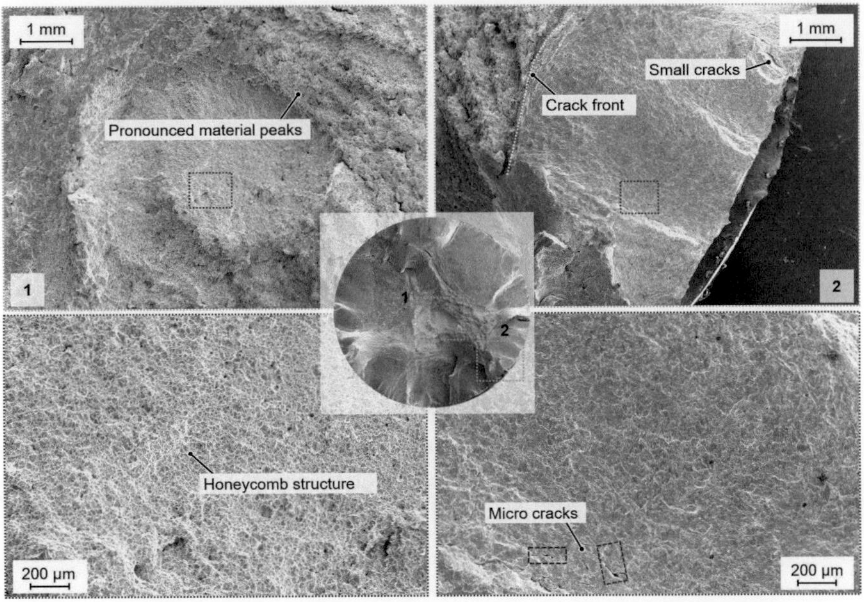

Figure 7.10: Detailed fractographs of AISI 304 specimen treated by means of MHP 10 setup
Detailaufnahmen der Bruchbilder von mittels MOH-Setup 10 bearbeiteten AISI 304 Proben

Figure 7.10 shows the fracture surface of a specimen machined using MHP-Setup 10. The area in the center of the specimen is clearly characterized by a structure with many small, rather flat honeycombs. Beach lines were not clearly determined. Larger material displacements and cracks in the center of the specimen as well as isolated small inclusions and pores are visible. Isolated small cracks run from the surface into the center of the workpiece.

7.3 Conclusion and validation of the research hypothesis

Fazit und Validierung der Forschungshypothese

Chapter 7 dealt with the research question **"What is the relationship between a knowledge-based adjustment of the surface integrity in electrodynamic MHP and the resulting fatigue strength?"** The answer to the question represented a deductive approach, with which the design model synthesized from the results of Chapter 4 to 6 should be examined.

For this purpose, fatigue tests were performed on a rotating bending test rig in this chapter. By means of a MHP test stand extension with an additional rotary axis, rotating bending specimens manufactured according to DIN50100 were processed using MHP. Thereby, roughness of up to Sa = 0.126 um and Sz = 3.93 um as well as surface hardness 475 $HV5$ were achieved. This corresponds approximately to the results of the flat samples from Chapter 6. Deviations result, among other things, from the different contact conditions due to higher pressures on rotationally symmetrical samples. Microstructure analyses confirmed the surface layer influenced by MHP, in which martensite was formed partially or homogeneously. By means of a rotating bending test rig, it was clearly shown that all three used MHP setups increased the fatigue strength of the AISI304 stainless steel under consideration here by a factor 1.15...1.21. Furthermore, the setups selected using the design model derived in Chapter 6 could confirm the conclusions. Decisive for an increased fatigue strength is a pronounced work hardening and high compressive residual stresses in the surface layer. In the considered area of low surface roughness, this plays a subordinate role for fatigue strength.

In summary, it can be stated that the design model derived from the inductive approach allows, under certain restrictions, an interpretation of the surface layer properties of austenitic stainless steel. The resulting surface layer properties after MHP have led to an improved fatigue strength. Accordingly, the research hypothesis **"By means of the defined energy input during robot-based electrodynamic machine hammer peening of austenitic stainless steel the surface integrity of the workpiece can specifically be adjusted and thus their fatigue strength can be improved"** formulated in Chapter 3 was validated.

8 Summary and outlook

Zusammenfassung und Ausblick

Anthropogenic global warming is the world's greatest challenge. The rising temperatures due to an increasing concentration of greenhouse gases, especially CO_2, endanger the global ecosystem. One approach to significantly reduce CO_2 emissions is the use of renewable energies in the energy industry. Mechanical components in energy generation plants are subject to particularly high loads due to the prevailing conditions. Austenitic stainless steels already used in some of these components are only used to a limited extent due to its partly insufficient mechanical strength and fatigue properties. However, the component properties can be significantly improved by mechanical surface layer treatment. Machine hammer peening (MHP) is an innovative process for surface layer treatment. Due to its process kinetics, it enables surfaces to be smoothed in a targeted manner and simultaneously high compressive residual stresses and work hardening to be introduced into the component surface layer. Both the MHP kinetics and the surface layer properties as well as the fatigue characteristics of austenitic stainless steel resulting from defined contact mechanisms have not yet been fully researched to the current state of the art. This results in a knowledge deficit regarding the cause-and-effect relationships between the MHP kinetics and the surface integrity and fatigue properties of austenitic stainless steel. Consequently, this work is based on the following research hypothesis: "**By means of the defined energy input during robot-based electrodynamic machine hammer peening of austenitic stainless steel the surface integrity of the workpiece can specifically be adjusted and thus their fatigue strength can be improved.** In order to validate the research hypothesis, this dissertation was structured according to the following research questions.

In a first step, the process kinetics of robot-guided MHP were analyzed in **Chapter 4**. The investigations were based on the following research question: *How do the process parameters of the robot-guided electrodynamic MHP influence the process kinetics and thus the resulting contact energy and how can the multiphysical system behavior be modeled to predict the process kinetics?* To answer this research question, first the effects of the dynamic robot behavior on the MHP kinetics were analyzed. Subsequently, an electrodynamic hammer system was extended metrologically, a complete system characterization and parameter study was performed, and the multiphysical system behavior was modeled using machine learning approaches. It was shown that the influences of the robot dynamics on the MHP kinetics and the machining result are negligible due to the sufficiently high robot stiffness. Furthermore, it was found that a sampling rate of $f_s \geq 50$ kHz enables precise data acquisition by the implemented fiber-optic distance and piezoelectric force sensors. The maximum hammering forces of the hammering system used were determined to be $F_{h,max}$ = 2.36 kN at a plunger stroke h = 1.0 mm and a hammering frequency f = 90 Hz. The degressively increasing course of the hammering force with increasing stroke as well as the antiproportionality between stroke and hammering frequency is due to the electrical power characteristic of the hammering system. The implemented distance sensor made it possible for the first time to record the stroke characteristics, which in part differ depending of the MHP parameters, and the resulting impact velocities of the plunger. It was determined that with increasing stroke or hammering frequency the contact energy increases non-linearly, partly degressively. Based on the available data and using machine-learning approaches, an experimental-mathematical as well as numerical model was developed to predict the contact energies as a function of the MHP process parameters.

In a second step, the contact mechanisms of individual impact events were investigated in **Chapter 5** using a drop tower test rig in analogy to the MHP. The procedure was based on the objective of answering the following research question: *How do defined contact energies in the electrodynamic MHP affect the contact conditions and the geometry of individual impacts and how do these influence the surface integrity of an austenitic stainless steel?* In order to answer this research question, the fundamental transferability of the results from the drop tower test rig to the MHP process was first demonstrated. It was shown experimentally that linear relationships as well as power laws describe the increase of indentation diameter, depth and volume with increasing contact energy. By means of the etching solution BERAHA II, it was also demonstrated that the metastable austenitic microstructure is partially transformed into the α'- martensite phase as a result of the energy input and that shear bands consisting of twins, stacking faults and ε-martensite are formed. An increased micro hardness confirmed this statement. It was shown that a sufficient indentation overlap, i.e. a sufficiently high energy density, is required for a homogeneous, deep-reaching microstructure transformation. Effective depths up to z = 1 mm could be recorded here. Supplementary numerical FE-simulations showed that the areas with pronounced transformation behavior correlate with the maximum plastic strains and strain rates. The plastic strain rates are in the MHP-relevant contact energy range $\dot{\varepsilon}_{\mathrm{pl}}$ = 2...6·10^3 s^{-1} and increase with smaller hammer head diameter due to higher HERTZ pressures. The significant influence of the plastic strain rate resulting from the contact energy on the transformation mechanisms in the material surface layer has been demonstrated.

In a third step, **Chapter 6** examined how the findings of individual impact events can be transferred to a real, two-dimensional MHP processing. The analyses in this chapter were therefore based on the following research question: *How do defined contact energies influence the surface integrity of austenitic stainless steels in the electrodynamic MHP?* By means of an experimental MHP parameter study, different surface layer states were set under variation of contact energies, energy densities and contact pressures. It was shown that the surface roughness of ground workpieces is mostly reduced and is mainly determined by the energy density and hammer head diameter. A significant increase in hardness of maximum $\Delta HV5$ = 182% and up to a surface layer depth z = 800...1,000 µm could be achieved. The cause-and-effect relationship between surface hardness and the MHP process parameters was also described by means of a linear correlation. In addition, it was found that softening processes occur in the surface layer when a certain energy threshold is exceeded. The distribution of the introduced high compressive residual stresses as a function of the MHP process parameters as well as the pronounced influence of the energy density on the residual stress values was discussed. The analyzed microstructure showed a clear, but partly not very homogeneous transformation behavior. Therefore, a sufficient energy input into the surface layer has to be realized by MHP. The austenitic phase fraction converted to martensite and the reduced grain size are in direct correlation to the energy density introduced. Using MHP, the dislocation density in the surface layer was increased and nanocrystals were generated. A sufficiently high energy density as well as contact pressure ensure a microstructure with homogeneous dislocation density and martensite transformation. Additional numerical FE-analyses using a dislocation density based material model have confirmed the increased dislocation density by MHP. The synthesis of the results from Chapter 4 to 6 into a MHP design model for austenitic stainless steels allows for future knowledge-based MHP processing.

In the fourth and last step in **Chapter 7**, the applicability of the design model on the one hand and the research hypothesis with the statement that the MHP leads to an increase in fatigue strength on the other hand were examined. The procedure was based on the following research question: *What is the relationship between a knowledge-based setting of the surface integrity in the electrodynamic MHP and the resulting fatigue strength?* By means of an MHP test rig extended by an additional external axis of rotation, cylindrical fatigue specimens were processed using selected MHP parameters and then tested for fatigue strength in a rotating bending test rig. The fatigue strength of the hammered workpieces was increased by a factor of 1.21 compared to the unhammered reference state. The findings regarding the surface integrity after the MHP, which were incorporated into the design model, were also confirmed. The targeted influence on the crack formation and crack propagation mechanisms could additionally be proven by a microscopic analysis of the fracture surfaces. The research hypothesis was validated accordingly.

The research objective **Knowledge-based explanatory model of the cause-and-effect relationships between the contact energies resulting from different process states in electrodynamic machine hammer peening and the workpiece surface layer of austenitic stainless steels as well as their effects on fatigue properties** was achieved.

Outlook

Based on the results of this dissertation, further questions arise in the field of MHP. It is necessary to investigate whether and, if so, under which restrictions the findings summarized in the design model developed can be transferred to other materials and heat treatment conditions. The underlying deviating material-physical mechanisms must be taken into account in a model extension. Furthermore, the explanatory approach described here for the surface layer and fatigue properties as a consequence of an energy-based MHP process design must be extended to machining under angular adjustment. Due to the already frequently addressed challenge regarding the MHP accessibility on complex components such as gears, impellers or compressor blades [TRAU16A], a non-orthogonal adjustment of the hammer system is likely. First investigations on a martensitic stainless steel X3CrNiMo13-4 showed that an angular adjustment of the MHP leads to reduced compressive residual stresses and higher roughness compared to an orthogonal adjustment, see Figure 8.1. A sliding or pulling MHP machining has a significant influence on the surface structure. Furthermore, it was shown that areas with property gradients can be achieved in MHP by a continuous stroke and thus energy reduction [MANN18B].

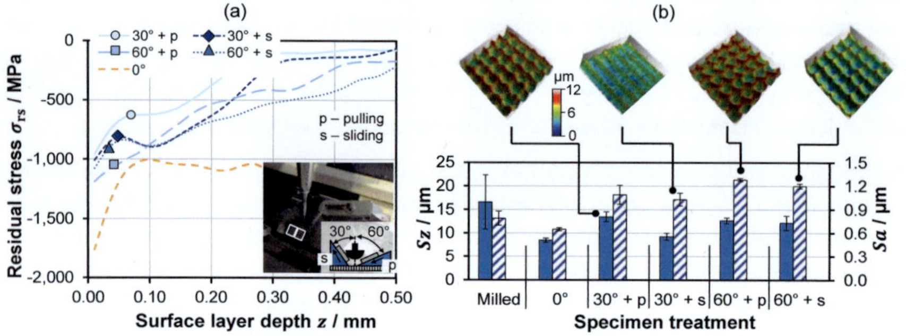

Figure 8.1: Influence of different machining directions and impact angles in MHP on residual stresses (a) and surface roughness (b) of X3CrNiMo13-4 stainless steel
Einfluss unterschiedlicher Bearbeitungsrichtungen und Auftreffwinkel auf die Eigenspannungen (a) und Oberflächenrauheit (b) von X3CrNiMo13-4 Edelstahl

In addition to the conventional process steps used for machining solid workpiece materials, such as forging, machining and subsequent or intermediate heat treatments, additive manufacturing processes are becoming increasingly important. However, the high economic potential of additive manufacturing processes is often offset by the challenges of an inferior surface integrity. A pronounced wavy surface morphology, inhomogeneous microstructures and tensile residual stresses, which can lead to workpiece distortion, are examples of this. At this point, the MHP offers the potential to specifically improve the surface layer properties in the form of intermediate or final-layer processing in the semi-warm or cold workpiece state. The use of MHP as a subsequent process step of laser deposition welding in hybrid machining centers or on industrial robots could be implemented by developing adapted tool holders. Preliminary investigations in collaboration with FRAUNHOFER IPT and FRAUNHOFER ILT on MHP final-layer machining of laser-welded IN718 powder and wire as well as Ti_6Al_4V wire have already demonstrated the possibilities of this hybrid process control. For the IN718 material, it was possible to significantly level the waviness, almost completely close the near-surface porosity, and achieve grain refinement by heat treatment after the MHP. Some of the results are shown in Figure 8.2. In the case of the Ti_6Al_4V material, the combined processing of MHP and downstream heat treatment enabled the setting of a bimodal microstructure, which is positive for the mechanical material properties.

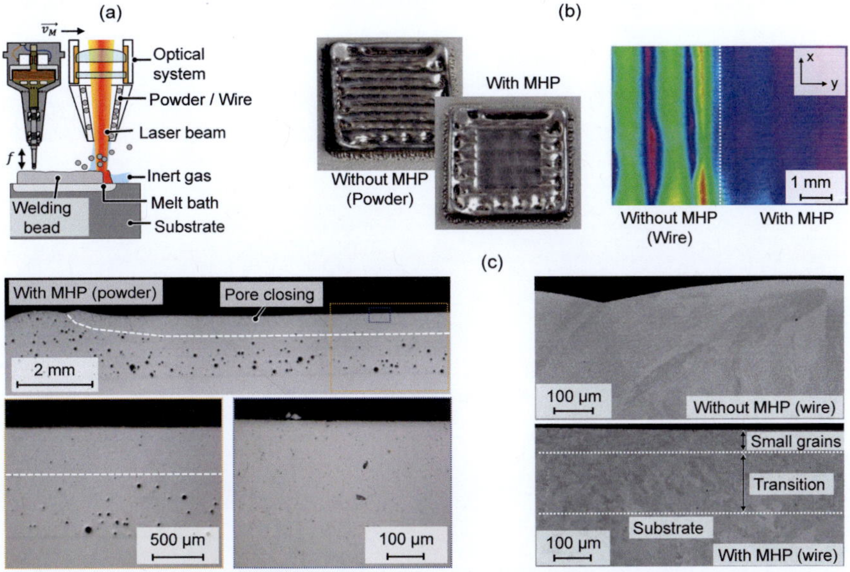

Figure 8.2: Schematic of the hybrid process of laser deposition welding and MHP (a) and analysis of surface morphology (b) and microstructure (c) of IN718 powder
Schema des hybriden Prozesses aus Laserauftragsschweißen und MOH (a) sowie Analyse der Oberflächenmorphologie (b) und Mikrostruktur (c) von IN718-Pulver

In addition to its use as a manufacturing process for solid components and tools, MHP can also be applied in sheet metal processing. For example, it is conceivable that MHP could be used as a substitution process for peen forming [MIAO11] of large structural components in the steel construction industry or flat outer skin and stringer components as used in the aerospace industry. The targeted, local application of compressive residual stresses could be used to curve flat sheets as well as to straighten curved sheets. The targeted positioning of the MHP offers the possibility of machining the components only exactly where adaptation of the surface layer properties is required. This would inevitably result in time and thus monetary savings in production.

Initial preliminary tests at the Laboratory for Machine Tools and Production Engineering (WZL) in collaboration with the Institute of Metal Forming (IBF) have already demonstrated the potential feasibility. The experimental setup (a) and the first results (b) are shown in Figure 8.3. Depending on the applied energy and energy density, sheets of the material AlMg3 (EN AW-5754) with thicknesses d_{B} = 1.0...1.5 mm were formed both concave and convex. Accordingly, the stresses introduced by means of MHP allowed the sheets to be formed as a function of the energy introduced.

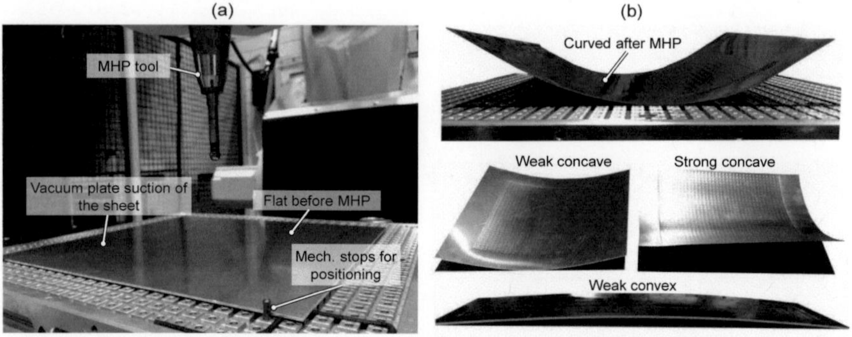

Figure 8.3: Test setup for MHP of sheet metal (a) and curved sheet metal (b)
Versuchsaufbau zum MOH von Blechen (a) und gekrümmte Bleche (b)

Zusammenfassung und Ausblick

Der antropogene Treibhauseffekt und daraus resultierende Klimawandel ist die größte Herausforderung der Menschheit. Die aufgrund einer zunehmenden Konzentration an Treibhausgasen, besonders CO_2, steigenden Temperaturen gefährden das weltweite Ökosystem. Ein Ansatz den CO_2-Ausstoß signifikant zu reduzieren besteht im Einsatz von erneuerbaren Energien im Bereich der Energiewirtschaft. Mechanische Bauteile in den Anlagen zur Energieumwandlung werden dabei aufgrund der vorherrschenden Bedingungen besonders stark belastet. Der bereits in einigen dieser Komponenten eingesetzte rostfreie austenitische Edelstahl kommt aufgrund seiner zum Teil unzureichenden mechanischen Festigkeits- und Ermüdungseigenschaften jedoch nur eingeschränkt zum Einsatz. Mittels einer mechanischen Randzonenbearbeitung können die Bauteileigenschaften jedoch signifikant verbessert werden. Das maschinelle Oberflächenhämmern (MOH) besitzt als innovatives Randzonenbearbeitungsverfahren aufgrund seiner Prozesskinetik die Möglichkeit, Oberflächen gezielt einzuglätten sowie hohe Druckeigenspannungen bei gleichzeitiger Kaltverfestigung in die Bauteilrandzone einzubringen. Sowohl die dem MOH zugrundeliegende Kinetik als auch die aus definierten Kontaktmechanismen resultierenden Randzonen- und Ermüdungseigenschaften von rostfreiem austenitischem Edelstahl sind zum aktuellen Stand der Technik noch unvollständig erforscht. Daraus lässt sich ein Wissensdefizit bzgl. der Ursache-Wirkungzusammenhänge zwischen der MOH-Kinetik und den resultierenden Randzonen- sowie den Ermüdungseigenschaften von rostfreiem austenitischen Edelstahl ableiten. Folglich liegt dieser Arbeit die folgende Forschungshypothese zu Grunde: **„Durch einen definierten Energieeintrag beim elektrodynamischen maschinellen Oberflächenhämmern von rostfreiem austenitischen Edelstahl kann die Randzonenintegrität von Werkstücken gezielt eingestellt und damit deren Ermüdungsfestigkeit verbessert werden“**. Um die Forschungshypothese zu validieren, wurde diese Dissertation entsprechend der folgenden Forschungsfragen gegliedert.

In einem ersten Schritt wurde dazu in **Kapitel 4** die Prozesskinetik des robotergeführten MOH analysiert. Die Untersuchungen basierten dabei auf folgender Forschungsfrage: *Wie beeinflussen die Prozessparameter beim robotergeführten elektrodynamischen MOH die Prozesskinetik und damit die resultierende Kontaktenergie und wie kann das multiphysikalische Systemverhalten zur Vorhersage der Prozesskinetik modelliert werden?* Zur Beantwortung der Forschungsfrage wurden zuerst die Auswirkungen des dynamischen Roboterverhaltens auf die MOH-Kinetik analysiert. Anschließend wurde ein elektrodynamisches Hämmersystem messtechnisch erweitert, eine vollumfängliche Systemcharakterisierung und Parameterstudie durchgeführt sowie das multiphysikalische Systemverhalten unter Zuhilfenahme maschineller Lernansätze modelliert. Es wurde gezeigt, dass die Einflüsse der Roboterdynamik auf die MOH-Kinetik und das Bearbeitungsergebnis aufgrund der ausreichend hohen Robotersteifigkeit vernachlässigbar sind. Weiterhin wurde festgestellt, dass eine Abtastrate von $f_s \geq 50$ kHz eine präzise Datenerfassung durch die implementierten faseroptischen Abstands- und piezoelektrischen Kraftsensoren ermöglicht. Die maximalen Hämmerkräfte des eingesetzten Hämmersystems wurden zu $F_{\mathrm{h,max}}$ = 2,36 kN bei einem Stößelhub h = 1,0 mm und einer Hämmerfrequenz f = 90 Hz bestimmt. Der degressiv ansteigende Verlauf der Hämmerkraft mit zunehmendem Hub sowie die Antiproportionalität zwischen Hub und Hämmerfrequenz sind dabei auf die elektrische Leistungscharakteristik des Hämmersystems zurückzuführen. Die implementierte Abstandssensorik ermöglichte erstmalig die Inline-Erfassung der aus den MOH-Parametern teils unterschiedlichen Hubverläufe und darauf resultierenden Aufprallgeschwindigkeiten des Hammerstößels. Es wurde ermittelt, dass mit steigendem Hub bzw. Hämmerfrequenz die Kontaktenergie nichtlinear, teils degressiv ansteigt. Aufbauend auf den erfassten Daten und unter Anwendung maschineller Lernansätze wurden sowohl ein empirisch-mathematisches als auch numerisches Modell zur Vorhersage der Kontaktenergien in Abhängigkeit der MOH-Prozessparameter entwickelt.

In einem zweiten Schritt wurden in **Kapitel 5** die Kontaktmechanismen einzelner Aufprallereignisse mittels eines Fallturmprüfstands in Analogie zum MOH untersucht. Das Vorgehen basierte dabei auf dem Ziel der Beantwortung folgender Forschungsfrage: *Wie beeinflussen definierte Kontaktenergien beim elektrodynamischen MOH die Kontaktbedingungen und die Geometrie von Einzeleinschlägen und wie wirken diese sich auf die Randzonenintegrität eines austenitischen rostfreien Edelstahls aus?* Zur Beantwortung der Forschungsfrage wurde zuerst die grundlegende Übertragbarkeit der Ergebnisse aus dem Fallturmprüfstand auf den MOH-Prozess nachgewiesen. Es wurde experimentell gezeigt, dass der Anstieg der Eindruckdurchmesser, -tiefe und -volumen mit steigender Kontaktenergie mithilfe von linearen Zusammenhängen als auch Potenzgesetzen beschrieben wird. Durch Anwendung der Ätzlösung BERAHA II wurde außerdem nachgewiesen, dass die metastabile austenitische Mikrostruktur infolge des Energieeintrags teilweise in die α'-Martensitphase umwandelt und Scherbänder bestehend aus Zwillingen, Stapelfehlern und ε-Martensit gebildet werden. Eine erhöhte Mikrohärte bestätigte diese Aussage. Dabei wurde gezeigt, dass eine ausreichende Eindrucküberlappung, also eine ausreichend hohe Energiedichte, für eine homogene, tief reichende Mikrostrukturumwandlung erforderlich ist. Wirktiefen bis z = 1 mm konnten hier erfasst werden. Durch ergänzende numerische FE-Simulationen wurde gezeigt, dass die Bereiche mit ausgeprägtem Umwandlungsverhalten mit den maximalen plastischen Dehnungen und Dehnraten korrelieren. Die plastischen Dehnraten liegen dabei im MOH-relevanten Kontaktenergiebereich bei $\dot{\varepsilon}_{\mathrm{pl}} = 2 \ldots 6 \cdot 10^3\ \mathrm{s}^{-1}$ und steigen mit kleinerem Hammerkopfdurchmesser aufgrund höherer HERTZSCHER Pressungen an. Der signifikante Einfluss der aus der Kontaktenergie resultierenden plastischen Dehnrate auf die Umwandlungsmechanismen in der Werkstoffrandzone konnte belegt werden.

In einem dritten Schritt wurde in **Kapitel 6** untersucht, wie sich die Erkenntnisse von einzelnen Aufprallereignissen auf eine reale, flächige MOH-Bearbeitung übertragen lassen. Die Analysen in diesem Kapitel basierten daher auf folgender Forschungsfrage: *Wie beeinflussen definierte Kontaktenergien die Randzonenintegrität von austenitischen Edelstählen beim elektrodynamischen MOH?* Mittels einer experimentellen

MOH-Parameterstudie wurden unter Variation der Kontaktenergien, Energiedichten und Kontaktpressungen unterschiedliche Randzonenzustände eingestellt. Es wurde gezeigt, dass die Oberflächenrauheit von geschliffenen Werkstücken zumeist reduziert und maßgeblich durch die Energiedichte und Hammerkopfdurchmesser bestimmt wird. Eine signifkante Härtesteigerung von maximal $\Delta HV5$ = 182 % und bis in eine Randzonentiefe z = 800...1000 µm konnte dabei erzielt werden. Die Ursache-Wirkungzusammenhänge zwischen der Oberflächenhärte und den MOH-Prozessparametern wurden außerdem mittels eines linearen Zusammenhangs beschrieben. Ergänzend wurde festgestellt, dass bei Überschreitung einer gewissen Energieschwelle Entfestigungsvorgänge in der Randzone auftreten. Die Verteilung der eingebrachten hohen Druckeigenspannungen in Abhängigkeit der MOH-Prozessparameter sowie der ausgeprägte Einfluss der Energiedichte auf die Eigenspannungsbeträge wurde diskutiert. Die analysierte Mikrostruktur zeigte ein deutliches, jedoch teils wenig homogenes Umwandlungsverhalten. Einen ausreichenden Energieeintrag in die Werkstückrandzone gilt es demnach durch das MOH zu realisieren. Der in Martensit umgewandelte austenitische Phasenanteil sowie die reduzierte Korngröße stehen dabei in direkter Korrelation zur eingebrachten Energiedichte. Ergänzende numerische FE-Analysen unter Einbindung eines versetzungsdichten-basierten Werkstoffmodells haben zusätzlich die erhöhte Versetzungsdichte durch das MOH bestätigt. Die Synthetisierung der Ergebnisse aus Kapitel 4 bis 6 in ein MOH-Gestaltungsmodell für austenitische rostfreie Edelstähle ermöglicht eine wissensbasierte MOH-Bearbeitung.

Im vierten und letzten Schritt in **Kapitel 7** wurde einerseits die Anwendbarkeit des Gestaltungsmodells, andererseits die Forschungshypothese mit der Aussage, dass das MOH zu einer Erhöhung der Dauerfestigkeit führt, überprüft. Dem Vorgehen lag folgende Forschungsfrage zugrunde: *Welcher Zusammenhang besteht zwischen einer wissensbasierten Einstellung der Randzonenintegrität beim elektrodynamischen MOH und der resultierenden Ermüdungsfestigkeit?* Mittels eines um eine zusätzliche externe Drehachse erweiterten MOH-Prüfstandes wurden zylindrische Ermüdungsproben unter Einsatz ausgewählter MOH-Parameter bearbeitet und anschließend in einem Biegeumlaufprüfstand auf ihre Dauerfestigkeit geprüft. Die Dauerfestigkeit der gehämmerten Werkstücke konnte dabei um den Faktor 1,21 gegenüber dem ungehämmerten Referenzzustand erhöht werden. Die in das Gestaltungsmodell eingeflossenen Erkenntnisse hinsichtlich der Randzonenintegrität nach dem MOH wurden ebenfalls bestätigt. Die gezielte Beeinflussung der Rissbildungs- und Rissausbreitungsmechanismen konnte zusätzlich durch eine mikroskopische Analyse der Bruchflächen gezeigt werden. Die Forschungshypothese wurde entsprechend validiert.

Das Forschungsziel **Wissensbasiertes Erklärungsmodells der Ursache-Wirkungzusammenhänge zwischen den aus verschiedenen Prozesszuständen resultierenden Kontaktenergien beim elektrodynamischen MOH und der Werkstückrandzone von austenitischen rostfreien Edelstählen sowie deren Auswirkungen auf die Ermüdungseigenschaften** wurde erreicht.

Ausblick

Basierend auf den Ergebnissen dieser Dissertation ergeben sich weitere Fragestellungen auf dem Gebiet des MOH. Es gilt zu untersuchen ob und ggf. unter welchen Restriktionen sich die Erkenntnisse, zusammengefasst in dem ausgearbeiteten Gestaltungsmodell, auf andere Werkstoffe und Wärmebehandlungszustände übertragen lassen. Die zugrundliegende abweichenden werkstoffphysikalischen Mechanismen müssen in einer Modellerweiterung berücksichtigt werden. Weiterhin gilt es den hier beschriebenen Erklärungsansatz der Randzonen- und Ermüdungseigenschaften als Folge einer energiebasierten MOH-Prozessauslegung auf eine Bearbeitung unter Winkelanstellung zu erweitern. Aufgrund der bereits häufig adressierten Herausforderung hinsichtlich der MOH-Zugänglichkeit an komplexen Bauteilen wie bspw. Zahnrädern, Impellern oder Verdichterschaufeln [TRAU16A], ist eine nichtorthogonale Anstellung des Hämmersystems wahrscheinlich. Erste Untersuchungen an einem martensitischen Edelstahl X3CrNiMo13-4 zeigten, dass eine Winkelanstellung beim MOH im Vergleich zu einer orthogonalen Anstellung zu reduzierten Druckeigenspannungen und höheren Rauheiten führt, siehe Abbildung 8.1. Eine schiebende bzw. ziehende MOH-Bearbeitung beeinflusst die Oberflächenstruktur dabei maßgeblich. Weiterhin wurde gezeigt, dass beim MOH Bereiche mit Eigenschaftsgradienten durch eine Hub- und somit Energiereduzierung erzielt werden können [MANN18B].

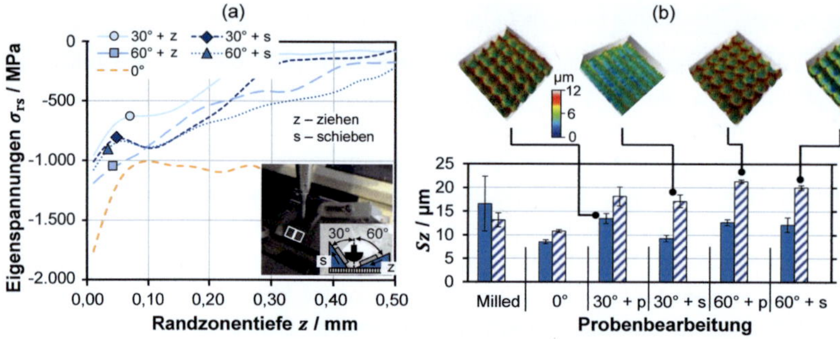

Abbildung 8.1: Einfluss unterschiedlicher Bearbeitungsrichtungen und Auftreffwinkel auf die Eigenspannungen (a) und Oberflächenrauheit (b) von X3CrNiMo13-4
Influence of different machining directions and impact angles in MHP on residual stresses (a) and surface roughness (b) of X3CrNiMo13-4

Neben den zur Bearbeitung von massiven Werkstückwerkstoffen konventionell eingesetzten Prozessschritten wie Schmieden, Zerspanung und nach- sowie zwischengeschalteter Wärmebehandlungen, gewinnen additive Fertigungsverfahren zunehmend an Bedeutung. Den hohen wirtschaftlichen Potenzialen der additiven Fertigungsverfahren stehen dabei jedoch häufig die Herausforderungen einer vergleichsweise minderwertigen Randzonenintegrität gegenüber. Eine ausgeprägte wellenförmige Oberflächenmorphologie, inhomogene Gefügestrukturen und Zugeigenspannungen, welche zu einem Werkstückverzug führen können, sind hier beispielhaft zu nennen. An dieser Stelle bietet das MOH das Potenzial in Form von Zwischen- oder Endlagenbearbeitung im halbwarmen bzw. kalten Werkstückzustand die Randzoneneigenschaften gezielt zu verbessern. Der Einsatz des MOH als dem Laserauftragsschweißen unmittelbar nachgelagerter Prozessschritt in hybriden Bearbeitungszentren oder auf Industrierobotern könnte dabei durch Entwicklung angepasster Werkzeugaufnahmen umgesetzt werden. Erste Voruntersuchungen in Zusammenarbeit mit dem FRAUNHOFER IPT und dem FRAUNHOFER ILT zum MOH-Endlagenbearbeitung von laseraufgeschweißtem IN718-Pulver und -draht sowie Ti_6Al_4V-Draht konnte bereits die vielfältigen Möglichkeiten dieser hybriden Prozessführung aufzeigen. Für den IN718-Werkstoff konnten dabei die Welligkeiten signifikant eingeebnet, die oberflächennahe Porosität nahezu vollständig geschlossen und eine Kornfeinung durch eine dem MOH nachgelagerte Wärmebehandlung erzielt werden. Ein Teil der Ergebnisse ist in Abbildung 8.2 dargestellt. Die kombinierte Bearbeitung aus MOH und nachgelagerter Wärmebehandlung ermöglichte im Falle des Ti6Al4V-Werkstoffs die Einstellung einer für die mechanischen Werkstoffeigenschaften positiven bimodalen Mikrostruktur.

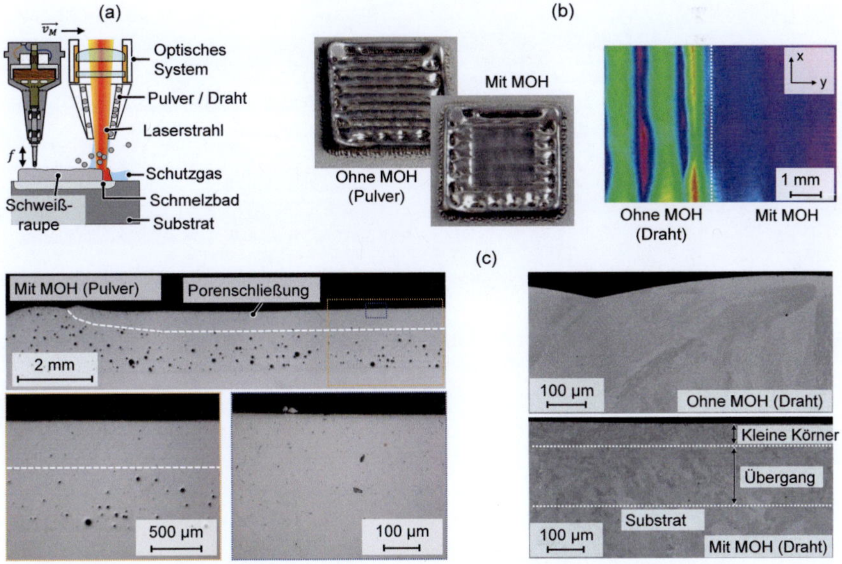

Abbildung 8.2: Schema des hybriden Prozesses aus Laserauftragsschweißen und MOH (a) sowie Analyse der Oberflächenmorphologie (b) und Mikrostruktur (c) von IN718-Pulver

Schematic of the hybrid process of laser deposition welding and MHP (a) and analysis of surface morphology (b) and microstructure (c) of IN718 powder

Darüber hinaus gibt es für das MOH neben dem Einsatz als Fertigungsverfahren im Bereich massiver Bauteil und Werkzeuge auch die Möglichkeit in der Blechbearbeitung Anwendung zu finden. So ist beispielweise vorstellbar, dass das MOH als Substitutionsprozess zum Richtstrahlen [MIAO11] großer Strukturbauteile in der Stahlbauindustrie oder flächiger Außenhaut- sowie Stringerbauteile, wie sie in der Luft- und

Raumfahrtindustrie verwendet werden, eingesetzt wird. Durch die gezielte, lokale Einbringung von Druckeigenspannungen könnten ebene Bleche gekrümmt sowie gekrümmte Bleche wieder gerichtet werden. Die gezielte Positionierung des MOH bietet die Möglichkeit die Bauteile nur genau dort zu bearbeiten, wo eine Anpassung der Randzoneneigenschaften gefordert wird. Daraus würden sich zwangsläufig zeitliche und somit monetäre Einsparungen in der Fertigung ergeben. Erste Vorversuche am WERKZEUGMASCHINENLABOR (WZL) in Zusammenarbeit mit dem INSTITUT FÜR BILDSAME FORMGEBUNG (IBF) haben die potenzielle Machbarkeit bereits aufgezeigt. Der Versuchsaufbau (a) und die ersten Ergebnisse (b) sind in Abbildung 8.3 dargestellt. In Abhängigkeit der eingebrachten Energie und Energiedichte wurden dabei Bleche aus dem Werkstoff AlMg3 (EN AW-5754) in Blechstärken d_B = 1,0…1,5 mm sowohl konkav als auch konvex ausgeformt. Durch die mittels MOH eingebrachten Spannungen konnten die Bleche in Abhängigkeit der eingebrachten Energie gezielt umgeformt werden.

(a) (b)

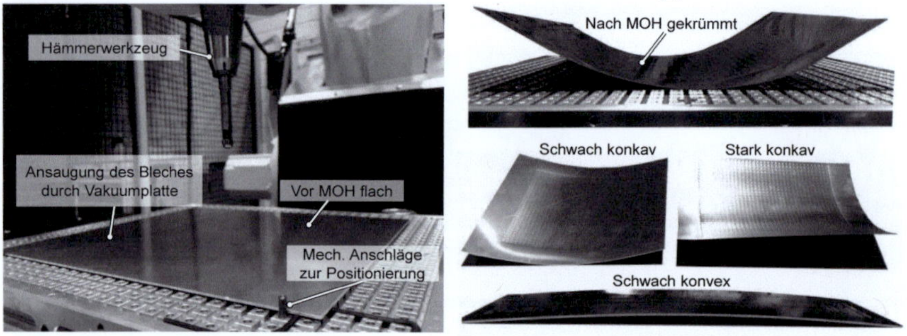

Abbildung 8.3: Versuchsaufbau zum MOH von Blechen (a) und gekrümmte Bleche (b)
Test setup for MHP of sheet metal (a) and curved sheet metal (b)

Bibliography

Literaturverzeichnis

[ADJA12] ADJASSOHO, B.; KOZESCHNIK, E.; LECHNER, C.; BLEICHER, F.; GOESSINGER, S.; BAUER, C.: *Induction of Residual Stresses and Increase of Surface Hardness by Machine Hammer Peening Technology*. In: Annals of DAAAM for 2012 & Proceedings of the 23rd International DAAAM Symposium, Vol. 23, 2012, No. 1, pp. 697–702

[ALHA82] AL-HASSANI, S. T. S.: *Mechanical aspects of residual stress development in shot peening*. In: Niku-Lari, A. (Ed.): First International Conference on Shot Peening. Oxford: Pergamon Press, 1982, pp. 583–602. ISBN: 9780080275994

[ALTE00] ALTENBERGER, I.: *Mikrostrukturelle Untersuchungen mechanisch randschicht-verfestigter Bereiche schwingend beanspruchter metallischer Werkstoffe*. PhD-Thesis. Universität Kassel, 2000. ISBN: 9783897920163

[ALTE03] ALTENBERGER, I.: *Alternative Mechanical Surface Treatments: Microstructures, Residual Stresses & Fatigue Behavior*. In: Wagner, L. (Ed.): Shot Peening. Weinheim, Cambridge: Wiley-VCH, 2003, pp. 419–434. ISBN: 9783527606580

[ALTE99] ALTENBERGER, I.; SCHOLTES, B.; MARTIN, U.; OETTEL, H.: *Cyclic deformation and near surface microstructures of shot peened or deep rolled austenitic stainless steel AISI 304*. In: Materials Science and Engineering: A, Vol. 264, 1999, pp. 1–16. DOI: 10.1016/S0921-5093(98)01121-6

[AMAN11] AMANOV, A.; PYOUN, Y.-S.; CHO, I.-S.; LEE, C.-S.; PARK, I.-G.: *The evaluation of the micro-tracks and micro-dimples on the tribological characteristics of thrust ball bearings*. In: Journal of Nanoscience and Nanotechnology, Vol. 11, 2011, No. 1, pp. 701–705. DOI: 10.1166/jnn.2011.3289

[AMAN19] AMANOV, A.; KARIMBAEV, R.; MALEKI, E.; UNAL, O.; PYUN, Y.-S.; AMANOV, T.: *Effect of combined shot peening and ultrasonic nanocrystal surface modification processes on the fatigue performance of AISI 304*. In: Surface and Coatings Technology, Vol. 358, 2019, pp. 695–705. DOI: 10.1016/j.surfcoat.2018.11.100

[ANGE54] ANGEL, T.: *Formation of Martensite in Austenitic Stainless Steel*. In: The Journal of the Iron and the Steel Institute (ISI), Vol. 177, 1954, pp. 165–174

[ARYA10] ARYAEI, A.; HASHEMNIA, K.; JAFARPUR, K.: *Experimental and numerical study of ball size effect on restitution coefficient in low velocity impacts*. In: International Journal of Impact Engineering, Vol. 37, 2010, No. 10, pp. 1037–1044. DOI: 10.1016/j.ijimpeng.2010.04.005

[BACA16] BACA, R. N.; REU, P. L.; ARAGON, D. S.; BRAKE, M. R. W.; VAN GOETHEM, D. J.; BEJARANO, M. V.; SUMALI, H.: *A Novel Experimental Method for Measuring Coefficients of Restitution. SAND2016-5693.* Sandia National Laboratories, Albuquerque (New Mexico, USA), 2016

[BAIN24] BAIN, E. C.; DUNKIRK, N. Y.: *The nature of Martensite*. In: Transactions of the Metallurgical Society of AIME, Vol. 70, 1924, pp. 25–45

[BALU13] BALUSAMY, T.; SANKARA NARAYANAN, T.S.N.; RAVICHANDRAN, K.; PARK, I. S.; LEE, M. H.: *Influence of surface mechanical attrition treatment (SMAT) on the corrosion behaviour of AISI 304 stainless steel.* In: Corrosion Science, Vol. 74, 2013, pp. 332–344. DOI: 10.1016/j.corsci.2013.04.056

[BARC65] BARCLAY, W. F.: *The Mechanisms of Deformation and Work Hardening in AISI Type 301 Stainless Steel.* In: ASTM International (Committee A-10) (Ed.): Advances in the Technology of Stainless Steels and Related Alloys. West Conshohocken, 1965. DOI: 10.1520/STP43727S

[BASQ10] BASQUIN, O. H.: *The Exponential Law of Endurance Tests.* In: Proceedings of the American Society for Testing and Materials (ASTM), Vol. 10, 1910, pp. 625–630

[BAUD77] BAUDRY, G.; PINEAU, A.: *Influence of strain-induced martensitic transformation on the low-cycle fatigue behavior of a stainless steel.* In: Materials Science and Engineering, Vol. 28, 1977, No. 2, pp. 229–242. DOI: 10.1016/0025-5416(77)90176-8

[BAUS86] BAUSCHINGER, J.: *Über die Veränderung der Elastizitätsgrenze und der Festigkeit des Eisens und Stahls durch Strecken und Quetschen, durch Erwärmen und Abkühlen und durch oftmals wiederholte Beanspruchung.* In: Mitteilungen aus dem Mechanisch-Technischen Laboratorium der K. Hochschule in München, Vol. 13, 1886, pp. 4–116

[BAYE89] BAYERLEIN, M.; CHRIST, H.-J.; MUGHRABI, H.: *Plasticity-induced martensitic transformation during cyclic deformation of AISI 304L stainless steel.* In: Materials Science and Engineering: A, Vol. 114, 1989, L11-L16. DOI: 10.1016/0921-5093(89)90871-X

[BEN19] BEN MOUSSA, N.; GHARBI, K.; CHAIEB, I.; BEN FREDJ, N.: *Improvement of AISI 304 austenitic stainless steel low-cycle fatigue life by initial and intermittent deep rolling.* In: The International Journal of Advanced Manufacturing Technology, Vol. 101, 2019, No. 1-4, pp. 435–449. DOI: 10.1007/s00170-018-2955-0

[BEND20] BENDER, B.; GERICKE, K.: *Pahl/Beitz Konstruktionslehre. Methoden und Anwendung erfolgreicher Produktentwicklung.* Vol. 9, Wiesbaden: Springer Vieweg, 2020. ISBN: 9783662573020

[BERG70] BERGSTRÖM, Y.: *A dislocation model for the stress-strain behaviour of polycrystalline α-Fe with special emphasis on the variation of the densities of mobile and immobile dislocations.* In: Materials Science and Engineering, Vol. 5, 1970, No. 4, pp. 193–200. DOI: 10.1016/0025-5416(70)90081-9

[BLEC01] BLECK, W.; FREHN, A.; JANKE, D.; FRANKE, A.: *Bedeutung der Umformtemperatur und -geschwindigkeit bei der Blechumformung austenitischer Edelstähle. Ergebnisse eines Vorhabens der industriellen Gemeinschaftsförderung (IGF).* Vol. 1, Hannover: Europäische Forschungsgemeinschaft für Blechverarbeitung e.V. (EFB), 2001. ISBN: 9783867762892

[BLEI12] BLEICHER, F.; LECHNER, C.; HABERSOHN, C.; KOZESCHNIK, E.; ADJASSOHO, B.; KAMINSKI, H.: *Mechanism of surface modification using machine hammer peening technology.* In: CIRP Annals, Vol. 61, 2012, No. 1, pp. 375–378. DOI: 10.1016/j.cirp.2012.03.139

[BLEI13] BLEICHER, F.; LECHNER, C.; HABERSOHN, C.; OBERMAIR, M.; HEINDL, F.; RODRIGUEZ RIPOLL, M.: *Improving the tribological characteristics of tool and mould surfaces by machine hammer peening*. In: CIRP Annals, Vol. 62, 2013, No. 1, pp. 239–242. DOI: 10.1016/j.cirp.2013.03.043

[BOLL70] BOLLING, G. F.; RICHMAN, R. H.: *The plastic deformation-transformation of paramagnetic f.c.c. FeNiC alloys*. In: Acta Metallurgica, Vol. 18, 1970, No. 6, pp. 673–681. DOI: 10.1016/0001-6160(70)90097-0

[BOYC01] BOYCE, B. L.; CHEN, X.; HUTCHINSON, J. W.; RITCHIE, R. O.: *The residual stress state due to a spherical hard-body impact*. In: Mechanics of Materials, Vol. 33, 2001, No. 8, pp. 441–454. DOI: 10.1016/S0167-6636(01)00064-3

[BRIN08] BRINKSMEIER, E.; GARBRECHT, M.; MEYER, D.: *Cold surface hardening*. In: CIRP Annals, Vol. 57, 2008, No. 1, pp. 541–544. DOI: 10.1016/j.cirp.2008.03.100

[BRIN14] BRINKSMEIER, E.; KLOCKE, F.; LUCCA, D. A.; SÖLTER, J.; MEYER, D.: *Process Signatures – A New Approach to Solve the Inverse Surface Integrity Problem in Machining Processes*. In: Procedia CIRP, Vol. 13, 2014, pp. 429–434. DOI: 10.1016/j.procir.2014.04.073

[BRIT16] BRITZ, D.; HEGETSCHWEILER, A.; ROBERTS, M.; MÜCKLICH, F.: *Reproducible Surface Contrasting and Orientation Correlation of Low-Carbon Steels by Time-Resolved Beraha Color Etching*. In: Materials Performance and Characterization, Vol. 5, 2016, No. 5, pp. 553–563. DOI: 10.1520/MPC20160067

[BRUN01] BRUNETTE, D. M.; TENGVALL, P.; TEXTOR, M.; THOMSEN, P.: *Titanium in Medicine. Material Science, Surface Science, Engineering, Biological Responses and Medical Applications,* Berlin, Heidelberg: Springer, 2001. ISBN: 9783642631191

[BUTL20] BUTLER, J. H.; MONTZKA, S. A.: *The NOAA Annual Greenhouse Gas Index (AGGI),* 2020. URL: https://www.esrl.noaa.gov/gmd/aggi/aggi.html

[BYUN04] BYUN, T.S.; HASHIMOTO, N.; FARRELL, K.: *Temperature dependence of strain hardening and plastic instability behaviors in austenitic stainless steels*. In: Acta Materialia, Vol. 52, 2004, No. 13, pp. 3889–3899. DOI: 10.1016/j.actamat.2004.05.003

[CELA17] CELADA-CASERO, C.; KOOIKER, H.; GROEN, M.; POST, J.; SAN-MARTIN, D.: *In-Situ Investigation of Strain-Induced Martensitic Transformation Kinetics in an Austenitic Stainless Steel by Inductive Measurements*. In: Metals, Vol. 7, 2017, No. 271, pp. 1–15. DOI: 10.3390/met7070271

[CHAN10] CHAN, H. L.; RUAN, H. H.; CHEN, A. Y.; LU, J.: *Optimization of the strain rate to achieve exceptional mechanical properties of 304 stainless steel using high speed ultrasonic surface mechanical attrition treatment*. In: Acta Materialia, Vol. 58, 2010, No. 15, pp. 5086–5096. DOI: 10.1016/j.actamat.2010.05.044

[CHAN18] CHANDRA, S.; SAMAL, M. K.; CHAVAN, V. M.; RAGHUNATHAN, S.: *Hierarchical multiscale modeling of plasticity in copper: From single crystals to polycrystalline aggregates*. In: International Journal of Plasticity, Vol. 101, 2018, pp. 188–212. DOI: 10.1016/j.ijplas.2017.10.014

[CHAT19] CHATTI, S.; LAPERRIÈRE, L.; REINHART, G.; TOLIO, T.: *CIRP Encyclopedia of Production Engineering,* Berlin, Heidelberg: Springer, 2019. ISBN: 9783662531198

[CHEN11A] CHEN, A. Y.; RUAN, H. H.; ZHANG, J. B.; LIU, X. R.; LU, J.: *Introducing a hierarchical structure for fabrication of a high performance steel*. In: Materials Chemistry and Physics, Vol. 129, 2011, No. 3, pp. 1096–1103. DOI: 10.1016/j.matchemphys.2011.05.068

[CHEN11B] CHEN, A. Y.; RUAN, H. H.; WANG, J.; CHAN, H. L.; WANG, Q.; LI, Q.; LU, J.: *The influence of strain rate on the microstructure transition of 304 stainless steel*. In: Acta Materialia, Vol. 59, 2011, No. 9, pp. 3697–3709. DOI: 10.1016/j.actamat.2011.03.005

[CHEN13] CHEN, T.; JOHN, H.; XU, J.; LU, Q.; HAWK, J.; LIU, X.: *Influence of surface modifications on pitting corrosion behavior of nickel-base alloy 718. Part 1: Effect of machine hammer peening*. In: Corrosion Science, Vol. 77, 2013, pp. 230–245. DOI: 10.1016/j.corsci.2013.08.007

[CHER10] CHERIF, A.; PYOUN, Y.; SCHOLTES, B.: *Effects of Ultrasonic Nanocrystal Surface Modification (UNSM) on Residual Stress State and Fatigue Strength of AISI 304*. In: Journal of Materials Engineering and Performance, Vol. 19, 2010, No. 2, pp. 282–286. DOI: 10.1007/s11665-009-9445-3

[CHOI97] CHOI, J.-Y.; JIN, W.: *Strain induced martensite formation and its effect on strain hardening behavior in the cold drawn 304 austenitic stainless steels*. In: Scripta Materialia, Vol. 36, 1997, No. 1, pp. 99–104. DOI: 10.1016/S1359-6462(96)00338-7

[CLAY06] CLAYTON, J. D.; MCDOWELL, D. L.; BAMMANN, D. J.: *Modeling dislocations and disclinations with finite micropolar elastoplasticity*. In: International Journal of Plasticity, Vol. 22, 2006, No. 2, pp. 210–256. DOI: 10.1016/j.ijplas.2004.12.001

[COFF53] COFFIN JR., L. F.: *A Study of the Effects of Cyclic Thermal Stresses on a Ductile Metal,* New York: Schenectady, 1953

[COST05] COSTA, H. L.: *Modification of Surface Topography: Manufacturing Methods and Applications*. PhD-Thesis. University of Cambridge, 2005

[COUR05] COURTNEY, T. H.: *Mechanical behavior of materials*. Vol. 2, Long Grove (Illinois, USA): Waveland Press, Inc., 2005. ISBN: 9781577664253

[CRUZ13] CRUZ FABIANO, A. L.: *Modelling of crystal plasticity and grain boundary migration of 304L steel at the mesoscopic scale*. PhD-Thesis. Ecole Nationale Supérieure des Mines de Paris, 2013

[DAI02] DAI, Q.-X.; WANG, A.-D.; CHENG, X.-N.; LUO, X.-M.: *Stacking fault energy of cryogenic austenitic steels*. In: Chinese Physics, Vol. 11, 2002, pp. 596–600. DOI: 10.1088/1009-1963/11/6/315

[DASH63] DASH, J.; OTTE, H. M.: *The martensite transformation in stainless steel*. In: Acta Metallurgica, Vol. 11, 1963, No. 10, pp. 1169–1178. DOI: 10.1016/0001-6160(63)90044-0

[DE06] DE, A. K.; SPEER, J. G.; MATLOCK, D. K.; MURDOCK, D. C.; MATAYA, M. C.; COMSTOCK, R. J.: *Deformation-induced phase transformation and strain hardening in type 304 austenitic stainless steel*. In: Metallurgical and Materials

Transactions A, Vol. 37, 2006, No. 6, pp. 1875–1886. DOI: 10.1007/s11661-006-0130-y

[DEMJ12] DEMJANENKO, D.: *Entwicklung eines elektrodynamischen, linearen Schwingaktors und das Verfahren zu dessen Steuerung und Regelung*. PhD-Thesis. Technische Universität Berlin, 2012. URI: 10.14279/depositonce-3312

[DEUT20] DEUTSCHE BUNDESREGIERUNG: *Ziele der Bundesregierung - Bis 2030 die Treibhausgase halbieren,* 2020. URL: https://www.bundesregierung.de/breg-de/themen/klimaschutz/klimaziele-und-sektoren-1669268

[DEW15] DEUTSCHE EDELSTAHLWERKE: *Werkstoffdatenblatt 42CrMo4 / 42CrMoS4 (1.7225 / 1.7227).* Schmolz + Bickenbach Group, 2015

[DEW16] DEUTSCHE EDELSTAHLWERKE: *Rost-, säure- und hitzebeständige Stähle.* Schmolz + Bickenbach Group, 2016

[DIN03] DEUTSCHES INSTITUT FÜR NORMUNG E. V.: *DIN 8580:2003-09: Fertigungsverfahren - Begriffe, Einteilung,* Berlin: Beuth Verlag GmbH. 2003-09

[DIN12] DEUTSCHES INSTITUT FÜR NORMUNG E. V.: *DIN EN ISO 25178-2: Geometrische Produktspezifikation (GPS) - Oberflächenbeschaffenheit: Flächenhaft - Teil 2: Begriffe und Oberflächen-Kenngrößen,* Berlin: Beuth Verlag GmbH. 2012-09

[DIN14A] DEUTSCHES INSTITUT FÜR NORMUNG E. V.: *DIN EN 10088-1:2014-12: Nichtrostende Stähle - Teil 1: Verzeichnis der nichtrostenden Stähle,* Berlin: Beuth Verlag GmbH. 2014-12

[DIN14B] DEUTSCHES INSTITUT FÜR NORMUNG E. V.: *DIN EN 10088-3:2014-12: Nichtrostende Stähle - Teil 3: Technische Lieferbedingungen für Halbzeug, Stäbe, Walzdraht, gezogenen Draht, Profile und Blankstahlerzeugnisse aus korrosionsbeständigen Stählen für allgemeine Verwendung,* Berlin: Beuth Verlag GmbH. 2014-12

[DIN16] DEUTSCHES INSTITUT FÜR NORMUNG E. V.: *DIN 50100:2016-12: Schwingfestigkeitsversuch - Durchführung und Auswertung von zyklischen Versuchen mit konstanter Lastamplitude für metallische Werkstoffproben und Bauteile,* Berlin: Beuth Verlag GmbH. 2016-12

[DIN18A] DEUTSCHES INSTITUT FÜR NORMUNG E. V.: *DIN EN ISO 6507-1: Metallische Werkstoffe - Härteprüfung nach Vickers - Teil 1: Prüfverfahren,* Berlin: Beuth Verlag GmbH. 2018-07

[DIN18B] DEUTSCHES INSTITUT FÜR NORMUNG E. V.: *DIN 50113:2018-12: Prüfung metallischer Werkstoffe - Umlaufbiegeversuch,* Berlin: Beuth Verlag GmbH. 2018-12

[DING11] DING, H.; SHEN, N.; SHIN, Y. C.: *Modeling of grain refinement in aluminum and copper subjected to cutting*. In: Computational Materials Science, Vol. 50, 2011, No. 10, pp. 3016–3025. DOI: 10.1016/j.commatsci.2011.05.020

[DING12A] DING, H.; SHEN, N.; SHIN, Y. C.: *Predictive modeling of grain refinement during multi-pass cold rolling*. In: Journal of Materials Processing Technology, Vol. 212, 2012, No. 5, pp. 1003–1013. DOI: 10.1016/j.jmatprotec.2011.12.005

[DING12B] DING, H.; SHIN, Y. C.: *Dislocation density-based modeling of subsurface grain refinement with laser-induced shock compression*. In: Computational Materials

Science, Vol. 53, 2012, No. 1, pp. 79–88. DOI: 10.1016/j.commatsci.2011.08.038

[DING14] DING, H.; SHIN, Y. C.: *Dislocation Density-Based Grain Refinement Modeling of Orthogonal Cutting of Titanium*. In: Journal of Manfacturing Science and Engineering, Vol. 136, 2014, No. 4, pp. 1–11. DOI: 10.1115/1.4027207

[DIXO48] DIXON, W. J.; MOOD, A. M.: *A Method for Obtaining and Analyzing Sensitivity Data*. In: Journal of the American Statistical Association, Vol. 43, 1948, No. 241, pp. 109–126. DOI: 10.1080/01621459.1948.10483254

[DOMK05] DOMKIN, K.: *Constitutive models based on dislocation density. Formulation and implementation into finite element codes*. PhD-Thesis. Luleå University of Technology, 2005. ISSN: 14021544

[DUEN84] DUENBOSTL, T.; OUDIN, T.; BASCHY, L.: *BASIC-Physikprogramme 2,* Wiesbaden: Vieweg+Teubner Verlag, 1984. ISBN: 9783322946737

[EICH53] EICHELMANN, G. H.; HULL, F. C.: *The Effect of Composition on the Temperature of Spontaneous Transformation of Austenite to Martensite in 18-8-Type Stainless Steel*. In: Transactions of the American Society for Metals (ASM), Vol. 45, 1953, pp. 77–104. DOI: 10.5781/JWJ.2018.36.2.5

[ESTR07] ESTRIN, Y.; KIM, H. S.: *Modelling microstructure evolution toward ultrafine crystallinity produced by severe plastic deformation*. In: Journal of Materials Science, Vol. 42, 2007, No. 5, pp. 1512–1516. DOI: 10.1007/s10853-006-1282-2

[ESTR98] ESTRIN, Y.; TÓTH, L. S.; MOLINARI, A.; BRÉCHET, Y.: *A dislocation-based model for all hardening stages in large strain deformation*. In: Acta Materialia, Vol. 46, 1998, No. 15, pp. 5509–5522. DOI: 10.1016/S1359-6454(98)00196-7

[FANG91] FANG, X. F.; DAHL, W.: *Strain hardening and transformation mechanism of deformation-induced martensite transformation in metastable austenitic stainless steels*. In: Materials Science and Engineering: A, Vol. 141, 1991, No. 2, pp. 189–198. DOI: 10.1016/0921-5093(91)90769-J

[FERR04] FERREIRA, P. J.; VANDER SANDE, J. B.; FORTES, M. A.; KYROLAINEN, A.: *Microstructure development during high-velocity deformation*. In: Metallurgical and Materials Transactions A, Vol. 35, 2004, No. 10, pp. 3091–3101. DOI: 10.1007/s11661-004-0054-3

[FIEL04] FIELD, J. E.; WALLEY, S. M.; PROUD, W. G.; GOLDREIN, H. T.; SIVIOUR, C. R.: *Review of experimental techniques for high rate deformation and shock studies*. In: International Journal of Impact Engineering, Vol. 30, 2004, No. 7, pp. 725–775. DOI: 10.1016/j.ijimpeng.2004.03.005

[FINN09] FINNEY, D. J.: *Probit analysis. A statistical treatment of the sigmoid response curve,* Cambridge: Cambridge University Press, 2009. ISBN: 9780521135900

[FOLL88] FOLLANSBEE, P. S.; KOCKS, U. F.: *A constitutive description of the deformation of copper based on the use of the mechanical threshold stress as an internal state variable*. In: Acta Metallurgica, Vol. 36, 1988, No. 1, pp. 81–93. DOI: 10.1016/0001-6160(88)90030-2

[FRAN50] FRANK, F. C.; READ, W. T.: *Multiplication Processes for Slow Moving Dislocations*. In: Physical Review, Vol. 79, 1950, No. 4, pp. 722–723. DOI: 10.1103/PhysRev.79.722

[FU15] FU, P.; JIANG, C.; WU, X.; ZHANG, Z.: *Surface Modification of 304 Steel Using Triple-Step Shot Peening*. In: Materials and Manufacturing Processes, Vol. 30, 2015, No. 6, pp. 693–698. DOI: 10.1080/10426914.2015.1004702

[FUCH83] FUCHSBAUER, B.: *Untersuchungen zur Schwingfestigkeitsoptimierung bauteilähnlicher Proben unterschiedlicher Größen durch Festwalzen*. PhD-Thesis. TU Darmstadt, 1983

[FÜRS19] FÜRST, T.: *Methodischer Festigkeitsnachweis von Bauteilen unter Berücksichtigung lokaler Härte und Einbauspiel am Beispiel formschlüssiger Kupplungen*. PhD-Thesis. TU Berlin, 2019. URI: 10.14279/depositonce-9181

[GATE13] GATEY, A. M.; HOSMANI, S. S.; SINGH, R.; SUWAS, S.: *Surface Engineering of Stainless Steels: Role of Surface Mechanical Attrition Treatment (SMAT)*. In: Advanced Materials Research, Vol. 794, 2013, pp. 238–247. DOI: 10.4028/www.scientific.net/AMR.794.238

[GEIG99] GEIGER, C.; KANZOW, C.: *Numerische Verfahren zur Lösung unrestringierter Optimierungsaufgaben,* Berlin, Heidelberg: Springer, 1999. ISBN: 9783642585821

[GELE07] GELETU, A.: *Solving optimization problems using the Matlab Optimization Toolbox - A tutorial.* TU Ilmenau, 2007

[GHAE17] GHAEDNIA, H.; WANG, X.; SAHA, S.; XU, Y.; SHARMA, A.; JACKSON, R. L.: *A Review of Elastic–Plastic Contact Mechanics*. In: Applied Mechanics Reviews, Vol. 69, 2017, No. 6. DOI: 10.1115/1.4038187

[GOLD15] GOLDSMITH, W.: *Impact: the theory and physical behaviour of colliding solids,* Mineola, NY: Dover Publications, 2015. ISBN: 0486420043

[GOTT07] GOTTSTEIN, G.: *Physikalische Grundlagen der Materialkunde*. Vol. 3, Berlin, Heidelberg: Springer-Verlag, 2007. ISBN: 9783540711049

[GOTT14] GOTTSTEIN, G.: *Materialwissenschaft und Werkstofftechnik. Physikalische Grundlagen*. Vol. 4, Berlin: Springer Vieweg, 2014. ISBN: 9783642366024

[GRÄN19] GRÄNING, T.: *Herstellung, Charakterisierung und Optimierung von austenitischen ODS Stählen*. PhD-Thesis. KIT Karlsruhe, 2019. DOI: 10.5445/KSP/1000075991

[GROC12] GROCHE, P.; ENGELS, M.; STEITZ, M.; MÜLLER, C.; SCHEIL, J.; HEILMAIER, M.: *Potential of mechanical surface treatment for mould and die production*. In: International Journal of Materials Research, Vol. 103, 2012, No. 6, pp. 783–789. DOI: 10.3139/146.110778

[GROS19] GROSDIDIER, T.; NOVELLI, M.: *Recent Developments in the Application of Surface Mechanical Attrition Treatments for Improved Gradient Structures: Processing Parameters and Surface Reactivity*. In: Materials Transactions, Vol. 60, 2019, No. 7, pp. 1344–1355. DOI: 10.2320/matertrans.MF201929

[GVR20] GRAND VIEW RESEARCH, INC.: *Stainless Steel Market Size, Share & Trends Analysis Report by Grade (200 Series, 300 Series, 400 Series, Duplex Series), by Product (Flat, Long), by Application (Building & Construction, Heavy Industry), by Region, and Segment Forecast, 2020-2027,* 2020. ISBN: 9781680389456

[HABE15] HABERSOHN, C.: *Analytische und Simulative Betrachtung eines Oberflächenhämmerprozesses.* PhD-Thesis. TU Wien, 2015. URI: http://hdl.handle.net/20.500.12708/2848

[HABS15] HABSCHIED, M.; GRAAF, B. de; KLUMPP, A.; SCHULZE, V.: *Fertigung und Eigenspannungen.* In: Journal of Heat Treatment and Materials (HTM), Vol. 70, 2015, No. 3, pp. 111–121. DOI: 10.3139/105.110261

[HACI09] HACINI, L.; VAN LÊ, N.; BOCHER, P.: *Evaluation of Residual Stresses Induced by Robotized Hammer Peening by the Contour Method.* In: Experimental Mechanics, Vol. 49, 2009, No. 6, pp. 775–783. DOI: 10.1007/s11340-008-9205-6

[HÄNS98] HÄNSEL, A.: *Nichtisothermes Werkstoffmodell für die FE-Simulation von Blechumformprozessen mit metastabilen austenitischen CrNi-Stählen.* PhD-Thesis. ETH Zürich, 1998. DOI: 10.3929/ethz-a-001947695

[HAUK06] HAUK, V.; BEHNKEN, H.: *Structural and residual stress analysis by nondestructive methods. Evaluation - application - assessment,* Amsterdam: Elsevier, 2006. ISBN: 9780444824769

[HECK82] HECKER, S. S.; STOUT, M. G.; STAUDHAMMER, K. P.; SMITH, J. L.: *Effects of Strain State and Strain Rate on Deformation-Induced Transformation in 304 Stainless Steel: Part I. Magnetic Measurements and Mechanical Behavior.* In: Metallurgical and Materials Transactions A, Vol. 13, 1982, No. 4, pp. 619–626. DOI: 10.1007/BF02644427

[HEIN04] HEINEMANN, G.: *Virtuelle Bestimmung des Verfestigungsverhaltens von Bändern und Blechen durch verformungsinduzierte Martensitbildung bei metastabilen rostfreien austenitischen Stählen.* PhD-Thesis. ETH Zurich, 2004. DOI: 10.3929/ETHZ-A-004906406

[HENN76] HENNESSY, D.; STECKEL, G.; ALTSTETTER, C.: *Phase transformation of stainless steel during fatigue.* In: Metallurgical and Materials Transactions A, Vol. 7, 1976, No. 3, pp. 415–424. DOI: 10.1007/BF02642838

[HERP00] HERPER, H. C.: *Ab-initio-Untersuchung magnetischer und struktureller Eigenschaften von 3d-Übergangsmetallen und ihren Legierungen.* PhD-Thesis. Universität Duisburg-Essen, 2000. URN: nbn:de:hbz:464-duett-05222001-0834193

[HERR08] HERRMANN, M.: *Entwurf, Berechnung und Regelung magnetischer Reluktanzaktoren.* PhD-Thesis. TU München, München, 2008. ISBN: 9783899638387

[HERT82] HERTZ, H.: *Über die Berührung fester elastischer Körper.* In: Journal für die reine und angewandte Mathematik, 1882, No. 92, pp. 156–171. DOI: 10.1515/crll.1882.92.156

[HOFF01] HOFFMANN, B.; VÖHRINGER, O.; MACHERAUCH, E.: *Effect of compressive plastic deformation on mean lattice strains, dislocation densities and flow stresses of martensitically hardened steels*. In: Materials Science and Engineering: A, Vol. 319-321, 2001, pp. 299–303. DOI: 10.1016/S0921-5093(01)00978-9

[HOFF11] HOFFMEISTER, J.; SCHULZE, V.; HESSERT, R.; KOENIG, G.: *Effects of the surface treatment on the measured diffraction peak width of Inconel 718*. In: International Scientific Committee for Shot Peening (Ed.): Proceedings of the 11th International Conference on Shot Peening, 2011, pp. 201–206

[HOLL45] HOLLOMON, J. R.: *Tensile deformation*. In: Transactions of American Institute of Mining, Metallurgical, and Petroleum Engineers (AIME), Vol. 162, 1945, pp. 268–277

[HOLZ98] HOLZAPFEL, H.; WICK, A.; SCHULZE, V.; VÖHRINGER, O.: *Einfluss der Kugelstrahlparameter auf die Randschichteigenschaften von 42CrMo4*. In: Härterei-technische Mitteilungen: HTM; Zeitschrift für Werkstoffe, Wärmebehandlung, Fertigung, Vol. 53, 1998, No. 3, pp. 155–163

[HONE81] HONEYCOMBE, R. W. K.: *Steels: Microstructure and properties,* London: Edward Arnold, 1981. ISBN: 9780713127935

[HORN15] HORNBOGEN, E.; WARLIMONT, H.: *Metalle. Struktur und Eigenschaften der Metalle und Legierungen*. Vol. 6, Berlin, Heidelberg: Springer Vieweg, 2015. ISBN: 9783662479513

[HOTZ19] HOTZ, H.; KIRSCH, B.; BECKER, S.; MÜLLER, R.; AURICH, J. C.: *Combination of cold drawing and cryogenic turning for modifying surface morphology of metastable austenitic AISI 347 steel*. In: Journal of Iron and Steel Research International, Vol. 26, 2019, No. 11, pp. 1188–1198. DOI: 10.1007/s42243-019-00306-x

[HU17] HU, D.; GAO, Y.; MENG, F.; SONG, J.; WANG, Y.; REN, M.; WANG, R.: *A unifying approach in simulating the shot peening process using a 3D random representative volume finite element model*. In: Chinese Journal of Aeronautics, Vol. 30, 2017, No. 4, pp. 1592–1602. DOI: 10.1016/j.cja.2016.11.005

[HUA06] HUA, M.; TAM, H. Y.; MA, H. Y.; MOK, C. K.: *Patterned PVD TiN spot coatings on M2 steel: Tribological behaviors under different sliding speeds*. In: Wear, Vol. 260, 2006, No. 11-12, pp. 1153–1165. DOI: 10.1016/j.wear.2005.07.014

[HUAN89] HUANG, G. L.; MATLOCK, D.; KRAUSS, G.: *Martensite formation, strain rate sensitivity, and deformation behavior of type 304 stainless steel sheet*. In: Metallurgical and Materials Transactions A, Vol. 20, 1989, No. 7, pp. 1239–1246. DOI: 10.1007/BF02647406

[HÜCK83] HÜCK, M.: *Ein verbessertes Verfahren für die Auswertung von Treppenstufenversuchen*. In: Materialwissenschaft und Werkstofftechnik, Vol. 14, 1983, No. 12, pp. 406–417. DOI: 10.1002/mawe.19830141207

[HUMP04] HUMPHREYS, F. J.; HATHERLY, M.: *Recrystallization and related annealing phenomena*. Vol. 2, Amsterdam: Elsevier, 2004. ISBN: 9780080441645

[IEA19] INTERNATIONAL ENERGY AGENCY: *World Energy Outlook 2019,* Paris: IEA Publications, 2019. ISBN: 9789264973008

[IEA20] INTERNATIONAL ENERGY AGENCY: *CO_2 Emissions from Fuel Combustion. Overview.* Inernational Energy Agency (IEA), Paris, 2020

[ISER08] INFORMATIONSSTELLE EDELSTAHL ROSTFREI: *Merkblatt 803 - Was ist nichtrostender Stahl?,* 2008

[ISSF16] INTERNATIONAL STAINLESS STEEL FORUM (ISSF): *Book of new applications 2016.* International Stainless Steel Forum (ISSF), 2016

[ISSF20] INTERNATIONAL STAINLESS STEEL FORUM (ISSF): *Stainless Steel in Figures 2020.* International Stainless Steel Forum (ISSF), 2020

[IWAM98] IWAMOTO, T.; TSUTA, T.; TOMITA, Y.: *Investigation on deformation mode dependence of strain-induced martensitic transformation in trip steels and modelling of transformation kinetics.* In: International Journal of Mechanical Sciences, Vol. 40, 1998, No. 2-3, pp. 173–182. DOI: 10.1016/S0020-7403(97)00047-7

[JOHN83] JOHNSON, G. R.; COOK, W. H.: *A Constitutive Model and Data for Metals Subjected to Large Strains, High Strain Rates, and High Temperatures.* In: Proceedings of the 7th International Symposium on Ballistics, The Hague, 19-21 April 1983, pp. 541–547

[JOHN85] JOHNSON, K. L.: *Contact mechanics,* Cambridge: Cambridge University Press, 1985. ISBN: 9781139171731

[KAI08] KAI, A.; SOYAMA, H.: *Visualization of the plastic deformation area beneath the surface of carbon steel induced by cavitation impact.* In: Scripta Materialia, Vol. 59, 2008, No. 3, pp. 272–275. DOI: 10.1016/j.scriptamat.2008.02.054

[KAIB01] KAIBYSHEV, O.A.: *Grain refinement in commercial alloys due to high plastic deformations and phase transformations.* In: Journal of Materials Processing Technology, Vol. 117, 2001, No. 3, pp. 300–306. DOI: 10.1016/S0924-0136(01)00784-1

[KALM11] KALMBACH, P.: *Effiziente Ableitungsbestimmung bei hochdimensionaler nichtlinearer Optimierung,* München: AVM, 2011. ISBN: 9783869241845

[KANA88] KANAZAWA, K.; YAMAGUCHI, K.; NISHIJIMA, S.: *Mapping of Low Cycle Fatigue Mechanisms at Elevated Temperatures for an Austenitic Stainless Steel.* In: Solomon, H.; Halford, G. (Ed.): Low Cycle Fatigue. Philadelphia (USA): ASTM International, 1988, 519–530. ISBN: 9780803109445

[KANO12] KANOU, S.; TAKAKUWA, O.; MANNAVA, S. R.; QIAN, D.; VASUDEVAN, V. K.; SOYAMA, H.: *Effect of the impact energy of various peening techniques on the induced plastic deformation region.* In: Journal of Materials Processing Technology, Vol. 212, 2012, No. 10, pp. 1998–2006. DOI: 10.1016/j.jmatprotec.2012.05.003

[KAZI13] KAZIMIERCZUK, M.: *High-Frequency Magnetic Components.* Vol. 2, Chichester (UK): John Wiley & Sons, Ltd, 2013. ISBN: 9780470748022

[KELL65] KELLY, P. M.: *The martensite transformation in steels with low stacking fault energy.* In: Acta Metallurgica, Vol. 13, 1965, No. 6, pp. 635–646. DOI: 10.1016/0001-6160(65)90126-4

[KIMU99] KIMURA, Y.; CHILDS, T.H.C.: *Surface asperity deformation under bulk plastic straining conditions.* In: International Journal of Mechanical Sciences, Vol. 41, 1999, No. 3, pp. 283–307. DOI: 10.1016/S0020-7403(98)00051-4

[KLAM10] KLAMSER, M.; LINDEN, P.; TSCHÜRTZ, A.; WIED, J.: *Cold forging device for metal surfaces, has lifting unit which is supported at its one end in bearing element, where impact head is arranged at opposite end of lifting unit. German Patent DE102009041720A1,* 2010

[KLES92] KLESNIL, M.; LUKÁS, P.: *Fatigue of metallic materials,* Amsterdam: Elsevier Science, 1992. ISBN: 9780444987235

[KLOC14] KLOCKE, F.; TRAUTH, D.; SCHONGEN, F.; SHIROBOKOV, A.: *Analysis of friction between stainless steel sheets and machine hammer peened structured tool surfaces: experimental and numerical investigation of the lubricated interaction gap.* In: Production Engineering, Vol. 8, 2014, No. 3, pp. 263–272. DOI: 10.1007/s11740-013-0519-6

[KLOC18] KLOCKE, F.; SHIROBOKOV, A.; HILD, R.; FEUERHACK, A.; TRAUTH, D.; MATTFELD, P.: *Friction and Wear Analysis of Deep Rolled and Vibrorolled Specimens in Lubricated Contact.* In: Key Engineering Materials, Vol. 767, 2018, pp. 93–100. DOI: 10.4028/www.scientific.net/KEM.767.93

[KLUM19] KLUMPP, A.: *Rissausbreitungs- und Rissöffnungsverhalten in mechanisch randnah und durchgreifend verfestigtem Inconel 718.* PhD-Thesis. KIT Karlsruhe, 2019. DOI: 10.5445/IR/1000098639

[KONG19] KONGTHEP, J.: *The Fatigue and Corrosion Fatigue Behavior of Austenitic Stainless Steel AISI 304 after Deep Rolling at Various Temperatures.* PhD-Thesis. Universität Kassel, 2019. DOI: 10.17170/kobra-20200114920

[KONG98] KONG, J.; CASSELL, A. M.; DAI, H.: *Chemical vapor deposition of methane for single-walled carbon nanotubes.* In: Chemical Physics Letters, Vol. 292, 1998, No. 4-6, pp. 567–574. DOI: 10.1016/S0009-2614(98)00745-3

[KORZ19] KORZYNSKI, M.: *Predicting the height of uneven surface after ball-peening machining.* In: International Journal of Mechanical Sciences, Vol. 150, 2019, pp. 617–624. DOI: 10.1016/j.ijmecsci.2018.10.066

[KRAL16] KRALL, S.; LECHNER, C.; NIRTL, M.; BLEICHER, F.: *Robot based machine hammer peening using an electromagnetic driven hammering device.* In: Proceedings of the 26th International Symposium on Intelligent Manufacturing and Automation, Vol. 1, 2016, pp. 633–640. DOI: 10.2507/26th.daaam.proceedings.086

[KRAL19] KRALL, S.: *Beeinflussung der Randschicht durch diskrete Mikroumformung.* PhD-Thesis. TU Wien, 2019

[KRAU10] KRAUER, J.: *Erweiterte Werkstoffmodelle zur Beschreibung des thermischen Umformverhaltens metastabiler Stähle.* PhD-Thesis. ETH Zurich, 2010. DOI: 10.3929/ETHZ-A-006234345

[KRUP07] KRUPP, U.: *Fatigue crack propagation in metals and alloys. Microstructural aspects and modelling concepts,* Weinheim: Wiley-VCH, 2007. ISBN: 9783527315376

[KURD30] KURDJUMOW, G.; SACHS, G.: *Über den Mechanismus der Stahlhärtung.* In: Zeitschrift für Physik, Vol. 64, 1930, No. 5-6, pp. 325–343. DOI: 10.1007/BF01397346

[KWEK18] KWEKU, D.; BISMARK, O.; MAXWELL, A.; DESMOND, K.; DANSO, K.; OTI-MENSAH, E.; QUACHIE, A.; ADORMAA, B.: *Greenhouse Effect: Greenhouse Gases and Their Impact on Global Warming*. In: Journal of Scientific Research and Reports, Vol. 17, 2018, No. 6, pp. 1–9. DOI: 10.9734/JSRR/2017/39630

[LAMP96] LAMPMAN, S. R.: *ASM Handbook: Volume 19 - Fatigue and fracture*. Vol. 3, Ohio (USA): ASM International, 1996. ISBN: 9780871703859

[LECH14] LECHNER, C.: *Oberflächenmodifikation unter Einsatz der Technologie des Schlagverdichtens (Machine Hammer Peenings)*. PhD-Thesis. TU Wien. URI: http://hdl.handle.net/20.500.12708/5636

[LEE00] LEE, W.-S.; LIN, C.-F.: *The morphologies and characteristics of impact-induced martensite in 304L stainless steel*. In: Scripta Materialia, Vol. 43, 2000, No. 8, pp. 777–782. DOI: 10.1016/S1359-6462(00)00487-5

[LEE04] LEE, Y.-L.; PAN, J.; HATHAWAY, R.; BARKEY, M.: *Fatigue Testing and Analysis: Theory and Practice,* Amsterdam: Elsevier Butterworth-Heinemann, 2004. ISBN: 9780750677196

[LEMI10] LEMIALE, V.; ESTRIN, Y.; KIM, H. S.; O'DONNELL, R.: *Grain refinement under high strain rate impact: A numerical approach*. In: Computational Materials Science, Vol. 48, 2010, No. 1, pp. 124–132. DOI: 10.1016/j.commatsci.2009.12.018

[LESY19] LESYK, D. A.; SOYAMA, H.; MORDYUK, B. N.; DZHEMELINSKYI, V. V.; MARTINEZ, S.; KHRIPTA, N. I.; LAMIKIZ, A.: *Mechanical Surface Treatments of AISI 304 Stainless Steel: Effects on Surface Microrelief, Residual Stress, and Microstructure*. In: Journal of Materials Engineering and Performance, Vol. 28, 2019, No. 9, pp. 5307–5322. DOI: 10.1007/s11665-019-04273-y

[LEVE44] LEVENBERG, K.: *A method for the solution of certain non-linear problems in least squares*. In: Quarterly of Applied Mathematics, Vol. 2, 1944, No. 2, pp. 164–168. DOI: 10.1090/QAM/10666

[LI19] LI, K.; CHEN, D.; CAO, P.; CAO, X.; SHIN, K.: *Effect of Plastic Deformation on Microstructure and Properties of 347 Austenite Steel*. In: Journal of Nanoscience and Nanotechnology, Vol. 19, 2019, No. 7, pp. 4078–4082. DOI: 10.1166/jnn.2019.16149

[LI20] LI, B.; ZHANG, S.; HU, R.; ZHANG, X.: *Dislocation density and grain size evolution in hard machining of H13 steel: numerical and experimental investigation*. In: Journal of Materials Research and Technology, Vol. 9, 2020, No. 3, pp. 4241–4254. DOI: 10.1016/j.jmrt.2020.02.051

[LIDA00] LIDA, K.: *Dent and Affected Layer Produced by Shot Peening*. In: The Shot Peener, Vol. 14, 2000, No. 3, pp. 4–7

[LIEN17] LIENERT, F.: *Bauteiloptimierung mittels Piezopeening - Auswirkungen auf den Randschichtzustand und die Schwingfestigkeit von 42CrMo4 V450*. PhD-Thesis. KIT Karlsruhe, 2017. DOI: 10.5445/IR/1000071646

[LIU01] LIU, G.; WANG, S. C.; LOU, X. F.; LU, J.; LU, K.: *Low carbon steel with nanostructured surface layer induced by high-energy shot peening*. In: Scripta Materialia, Vol. 44, 2001, pp. 1791–1795. DOI: 10.1016/S1359-6462(01)00738-2

[LIVI77] LIVITSANOS, C. P.; THOMSON, P. F.: *The effect of temperature and deformation rate on transformation-dependent ductility of a metastable austenitic stainless steel.* In: Materials Science and Engineering, Vol. 30, 1977, No. 2, pp. 93–98. DOI: 10.1016/0025-5416(77)90216-6

[LLEW97] LLEWELLYN, D. T.: *Work hardening effects in austenitic stainless steels*. In: Materials Science and Technology, Vol. 13, 1997, No. 5, pp. 389–400. DOI: 10.1179/mst.1997.13.5.389

[LÖCK07] LÖCKER, C.: *Working, smoothing and cold hardening of the surface of tools, machine parts or other parts comprises hammering the surface with an electromagnetically controlled hammer head. German Patent DE102006033004A1,* 2007

[LÖCK12] LÖCKER, C.: *Method and device for the technique of cold microforging any freely formed 3-D surfaces: United States Patent US 8166793 B,* 2012

[LOWE17] LOWE, D. B.: *Das Chemiebuch. Vom Schießpulver bis zum Graphen, 250 Meilensteine in der Geschichte der Chemie,* Kerkdriel (Niederlande): Librero IBP, 2017. ISBN: 9789089988324

[LU04] LU, K.; LU, J.: *Nanostructured surface layer on metallic materials induced by surface mechanical attrition treatment*. In: Materials Science and Engineering: A, Vol. 375-377, 2004, pp. 38–45. DOI: 10.1016/j.msea.2003.10.261

[LU16] LU, J.; HULTMAN, L.; HOLMSTRÖM, E.; ANTONSSON, K. H.; GREHK, M.; LI, W.; VITOS, L.; GOLPAYEGANI, A.: *Stacking fault energies in austenitic stainless steels*. In: Acta Materialia, Vol. 111, 2016, pp. 39–46. DOI: 10.1016/j.actamat.2016.03.042

[LU92] LU, H.; SCHOLTES, B.; MACHERAUCH, E.: *Randschichtzustände von normalisiertem und vergütetem 42CrMo4 nach konsekutiven Kugelstrahl- und Festwalzbehandlungen*. In: Materials Science and Engineering Technology, Vol. 23, 1992, No. 11, pp. 388–394. DOI: 10.1002/mawe.19920231109

[MACH85] MACHERAUCH, E.; WOHLFAHRT, H.: *Eigenspannungen und Ermüdung*. In: Munz, D. (Ed.): Ermüdungsverhalten metallischer Werkstoffe. Oberursel: DGM-Informationsgesellschaft Verlag, 1985, pp. 237–283

[MADE06] MADER, S.: *Festwalzen von Fan- und Verdichterschaufeln*. PhD-Thesis. RWTH Aachen University, 2006

[MAEN77] MAENNING, W. W.: *Das Abgrenzungsverfahren, eine kostsparende Methode zur Ermittlung von Schwingfestigkeitswerten*. In: Materialprüfung, Vol. 19, 1977, No. 8, pp. 280–289

[MANN17] MANNENS, R.; DELFORNO, A.; TRAUTH, D.; FEUERHACK, A.; KITTEL, M.; KLOCKE, F.: *Analysis of surface defects on industrial casting tools for automotive applications after machine hammer peening*. In: Procedia Engineering, Vol. 207, 2017, pp. 1141–1146. DOI: 10.1016/j.proeng.2017.10.1139

[MANN18A] MANNENS, R.; TRAUTH, D.; FALKER, J.; BRECHER, C.; KLOCKE, F.: *Analysis of spindle bearing load with regard to the false brinelling effect caused by machine hammer peening*. In: The International Journal of Advanced Manufacturing

Technology, Vol. 95, 2018, No. 9-12, pp. 3969–3976. DOI: 10.1007/s00170-017-1523-3

[MANN18B] MANNENS, R.; TRAUTH, D.; MATTFELD, P.; KLOCKE, F.: *Influence of Impact Force, Impact Angle, and Stroke Length in Machine Hammer Peening on the Surface Integrity of the Stainless Steel X3CrNiMo13-4*. In: Procedia CIRP, Vol. 71, 2018, pp. 166–171. DOI: 10.1016/j.procir.2018.05.091

[MANN20] MANNENS, R.; UHLMANN, L.; LAMBERS, F.; FEUERHACK, A.; BERGS, T.: *Surface Integrity of AISI 52100 Bearing Steel after Robot-Based Machine Hammer Peening*. In: Journal of Manufacturing and Materials Processing, Vol. 4(2), 2020, No. 61, pp. 1–17. DOI: 10.3390/jmmp4020061

[MANS53] MANSON, S. S.: *Technical Note 2933: Behaviour of material under conditions of thermal stress.* National Advisory Comittee for Aeronautics (NACA), Cleveland (Ohio, USA), 1953

[MARQ63] MARQUARDT, D. W.: *An Algorithm for Least-Squares Estimation of Nonlinear Parameters.* In: Journal of the Society for Industrial and Applied Mathematics, Vol. 11, 1963, No. 2, pp. 431–441. DOI: 10.1137/0111030

[MART80] MARTIN, P.: *Beitrag zur Ermittlung der Einflussgrößen beim Kugelstrahlen durch Einzelkornversuche*. PhD-Thesis. Universität der Bundeswehr Hamburg, 1980

[MECK81] MECKING, H.; KOCKS, U. F.: *Kinetics of flow and strain-hardening*. In: Acta Metallurgica, Vol. 29, 1981, No. 11, pp. 1865–1875. DOI: 10.1016/0001-6160(81)90112-7

[MESS14] MESSÉ, O. M. D. M.; STEKOVIC, S.; HARDY, M. C.; RAE, C. M. F.: *Characterization of Plastic Deformation Induced by Shot-Peening in a Ni-Base Superalloy.* In: The Journal of The Minerals, Metals & Materials Society (JOM), Vol. 66, 2014, No. 12, pp. 2502–2515. DOI: 10.1007/s11837-014-1184-8

[MEYE16] MEYER, D.; KÄMMLER, J.: *Surface Integrity of AISI 4140 After Deep Rolling with Varied External and Internal Loads*. In: Procedia CIRP, Vol. 45, 2016, pp. 363–366. DOI: 10.1016/j.procir.2016.02.356

[MIAO11] MIAO, H. Y.; LAROSE, S.; PERRON, C.; LÉVESQUE, M.: *Numerical simulation of the stress peen forming process and experimental validation*. In: Advances in Engineering Software, Vol. 42, 2011, No. 11, pp. 963–975. DOI: 10.1016/j.advengsoft.2011.05.025

[MONA18] MONAVARI, M.; ZAISER, M.: *Annihilation and sources in continuum dislocation dynamics*. In: Materials Theory, Vol. 2, 2018, No. 1, pp. 1–30. DOI: 10.1186/s41313-018-0010-z

[MONT11] MONTAINE, M.; HECKEL, M.; KRUELLE, C.; SCHWAGER, T.; PÖSCHEL, T.: *Coefficient of restitution as a fluctuating quantity*. In: Physical review E: Statistical, nonlinear, and soft matter physics, Vol. 84, 2011, No. 4(1), 041306:1-5. DOI: 10.1103/PhysRevE.84.041306

[MORR65] MORROW, J.: *Cyclic Plastic Strain Energy and Fatigue of Metals*. In: ASTM Special technical publication, Vol. 378, 1965, 45-87. DOI: 10.1520/STP43764S

[MOTA19] MOTAMAN, S.; PRAHL, U.: *Microstructural constitutive model for polycrystal viscoplasticity in cold and warm regimes based on continuum dislocation dynamics*. In: Journal of the Mechanics and Physics of Solids, Vol. 122, 2019, pp. 205–243. DOI: 10.1016/j.jmps.2018.09.002

[MOUS15] MOUSSA, C.; BERNACKI, M.; BESNARD, R.; BOZZOLO, N.: *About quantitative EBSD analysis of deformation and recovery substructures in pure Tantalum*. In: IOP Conference Series: Materials Science and Engineering, Vol. 89, 2015, p. 12038–12038. DOI: 10.1088/1757-899X/89/1/012038

[MUGH15] MUGHRABI, H.: *Microstructural mechanisms of cyclic deformation, fatigue crack initiation and early crack growth*. In: Philosophical Transactions A: Mathematical, physical, and engineering sciences, Vol. 373, 2015, No. 2038, pp. 1–21. DOI: 10.1098/rsta.2014.0132

[MÜLL11] MÜLLER-BOLLENHAGEN, C.: *Verformungsinduzierte Martensitbildung bei mehrstufiger Umformung und deren Nutzung zur Optimierung der HCF- und VHCF-Eigenschaften von austenitischem Edelstahlblech*. PhD-Thesis. Universität Siegen, 2011. ISSN: 21935114

[MÜLL12] MÜLLER, C.; HINKELMANN, K.; WÄCHTER, M.; MASENDORF, R.; ESDERTS, A.: *Zur Wiederverwendung von Durchläufern im Treppenstufenversuch*. In: Materials Testing, Vol. 54, 2012, No. 11-12, pp. 786–792. DOI: 10.3139/120.110390

[MUÑO17] MUÑOZ-CUBILLOS, J.; CORONADO, J. J.; RODRÍGUEZ, S. A.: *Deep rolling effect on fatigue behavior of austenitic stainless steels*. In: International Journal of Fatigue, Vol. 95, 2017, pp. 120–131. DOI: 10.1016/j.ijfatigue.2016.10.008

[MURR82] MURR, L. E.; STAUDHAMMER, K. P.; HECKER, S. S.: *Effects of Strain State and Strain Rate on Deformation-Induced Transformation in 304 Stainless Steel: Part II. Microstructural Study*. In: Metallurgical and Materials Transactions A, Vol. 13, 1982, No. 4, pp. 627–635. DOI: 10.1007/BF02644428

[NETO20] NETO, L.; WILLIAMS, S.; DING, J.; HÖNNIGE, J.; MARTINA, F.: *Mechanical Properties Enhancement of Additive Manufactured Ti-6Al-4V by Machine Hammer Peening*. In: Itoh, S.; Shukla, S. (Ed.): Advanced surface enhancement. Singapore: Springer, 2020, pp. 121–132. ISBN: 9789811500534

[NICH06] NICHOLAS, T.: *High cycle fatigue: A mechanics of materials perspective*. Vol. 1, Amsterdam: Elsevier Science, 2006. ISBN: 9780080446912

[NIKI07] NIKITIN, I.: *Zur Verbesserung des Ermüdungsverhaltens des austenitischen Stahls X5CrNi18-10 im Temperaturbereich 25-600 °C durch mechanische Randschichtverfestigungsverfahren*. PhD-Thesis. Universität Kassel, 2007. ISBN: 9783899583557

[NISH78] NISHIYAMA, Z.: *Martensitic Transformation,* Oxford: Elsevier Science, 1978. ISBN: 9780125198509

[NOHA77] NOHARA, K.; ONO, Y.; OHASHI, N.: *Composition and Grain Size Dependencies of Strain-induced Martensitic Transformation in Metastable Austenitic Stainless Steels*. In: The Journal of the Iron and the Steel Institute (ISI), Vol. 63, 1977, No. 5, pp. 212–222. DOI: 10.2355/tetsutohagane1955.63.5_772

[OLLE76] OLLENDORFF, F.: *Induktivität, Widerstand und Eigenkraft der Tauchspule elektrodynamischer Lautsprecher*. In: Archiv für Elektrotechnik, Vol. 58, 1976, No. 3, pp. 151–159. DOI: 10.1007/BF01579716

[OLSO72] OLSON, G. B.; COHEN, M.: *A mechanism for the strain-induced nucleation of martensitic transformations*. In: Journal of the Less Common Metals, Vol. 28, 1972, No. 1, pp. 107–118. DOI: 10.1016/0022-5088(72)90173-7

[OLSO75] OLSON, G. B.; COHEN, M.: *Kinetics of strain-induced martensitic nucleation*. In: Metallurgical and Materials Transactions A, Vol. 6, 1975, No. 4, pp. 791–795. DOI: 10.1007/BF02672301

[OLUG20] OLUGBADE, T. O.; LU, J.: *Literature review on the mechanical properties of materials after surface mechanical attrition treatment (SMAT)*. In: Nano Materials Science, Vol. 2, 2020, No. 1, pp. 3–31. DOI: 10.1016/j.nanoms.2020.04.002

[OROW34] OROWAN, E.: *Zur Kristallplastizität. III - Über den Mechanismus des Gleitvorgangs*. In: Z. Physik (Zeitschrift für Physik), Vol. 89, 1934, No. 9-10, pp. 605–613. DOI: 10.1007/BF01341478

[OSTE16] OSTERTAG, A.; ZENK, S.: *Clamping device for influencing workpieces and associated method. German Patent DE102015203487A1,* 2016

[PAPA17] PAPANIKOLAOU, S.; CUI, Y.; GHONIEM, N.: *Avalanches and plastic flow in crystal plasticity: an overview*. In: Modelling and Simulation in Materials Science and Engineering, Vol. 26, 2017, No. 1, 013001:1-55

[PATE53] PATEL, J. R.; COHEN, M.: *Criterion for the action of applied stress in the martensitic transformation*. In: Acta Metallurgica, Vol. 1, 1953, No. 5, pp. 531–538. DOI: 10.1016/0001-6160(53)90083-2

[PATI17] PATIL, D.; HIGGS, C. F.: *A coefficient of restitution model for sphere–plate elastoplastic impact with flexural vibrations*. In: Nonlinear Dynamics, Vol. 88, 2017, No. 3, pp. 1817–1832. DOI: 10.1007/s11071-017-3346-z

[PEYR00] PEYRE, P.; SCHERPEREEL, X.; BERTHE, L.; CARBONI, C.; FABBRO, R.; BÉRANGER, G.; LEMAITRE, C.: *Surface modifications induced in 316L steel by laser peening and shot-peening. Influence on pitting corrosion resistance*. In: Materials Science and Engineering: A, Vol. 280, 2000, No. 2, pp. 294–302. DOI: 10.1016/S0921-5093(99)00698-X

[PICK13] PICKERD, V.: *Optimisation and Validation of the ARAMIS Digital Image Correlation System for use in Large-scale High Strain-rate Events*. Commonwealth of Australia - Maritime Division, Victoria (Australia), 2013

[PINT12] PINTAUDE, G.; HOECHELE, A. R.; CIPRIANO, G. L.: *Relation between strain hardening exponent of metals and residual profiles of deep spherical indentation*. In: Materials Science and Technology, Vol. 28, 2012, No. 9-10, pp. 1051–1054. DOI: 10.1179/1743284711Y.0000000107

[PIPP06] PIPPAN, R.; WETSCHER, F.; HAFOK, M.; VORHAUER, A.; SABIROV, I.: *The Limits of Refinement by Severe Plastic Deformation*. In: Advanced Engineering Materials, Vol. 8, 2006, No. 11, pp. 1046–1056. DOI: 10.1002/adem.200600133

[POPO09] POPOV, V. L.: *Kontaktmechanik und Reibung. Ein Lehr- und Anwendungsbuch von der Nanotribologie bis zur numerischen Simulation,* Berlin, Heidelberg: Springer, 2009. ISBN: 9783540888369

[POPO18] POPOV, V. L.; HEß, M.; WILLERT, E.: *Handbuch der Kontaktmechanik,* Berlin, Heidelberg: Springer, 2018. ISBN: 9783662530108

[POWE58] POWELL, G. W.; MARSHALL, E. R.; BACKOFEN, W. A.: *Strain Hardening of Austenitic Stainless Steel.* In: Transactions of the American Society for Metals (ASM), Vol. 50, 1958, No. 1, pp. 478–497

[PRAM15] PRAMANIK, A.; BASAK, A. K.: *Stainless steel. Microstructure, mechanical properties and methods of application,* Hauppauge (New York, USA): Nova Science Publishers Inc, 2015. ISBN: 9781634820813

[RADA07] RADAJ, D.; VORMWALD, M.: *Ermüdungsfestigkeit. Grundlagen für Ingenieure.* Vol. 3, Berlin, Heidelberg: Springer-Verlag, 2007. ISBN: 9783540714590

[RAMA06] RAMASWAMY, V.; SCHWARZKOPF, M. D.; RANDEL, W. J.; SANTER, B. D.; SODEN, B. J.; STENCHIKOV, G. L.: *Anthropogenic and natural influences in the evolution of lower stratospheric cooling.* In: Science, Vol. 311, 2006, No. 5764, pp. 1138–1141. DOI: 10.1126/science.1122587

[RANS53] RANSOM, J. T.; MEHL, R. F.: *The Statistical Nature of the Fatigue Properties of SAE 4340 Steel Forgings.* In: Baker, M. (Ed.): Symposium on Fatigue with Emphasis on Statistical Approach-II. West Conshohocken (USA): American Society for Testing & Materials, 1953, 3-22. ISBN: 9780803162600

[RASC12] Rasch, M.; Mainz, J.: *100 Jahre nichtrostender Stahl. Historisches und Aktuelles*, Essen: Klartext-Verlag, 2012. ISBN: 9783837500950

[REBH15] REBHAN, E.: *Theoretische Physik: Elektrodynamik.* Vol. 1, Berlin: Springer, 2015. ISBN: 9783662462942

[RHOD77] RHODES, C. G.; THOMPSON, A. W.: *The composition dependence of stacking fault energy in austenitic stainless steels.* In: Metallurgical and Materials Transactions A, Vol. 8, 1977, No. 12, pp. 1901–1906. DOI: 10.1007/BF02646563

[RIED10] RIEDNER, S.: *Höchstfeste nichtrostende austenitische CrMn-Stähle mit (C+N).* PhD-Thesis. Ruhr-Universität Bochum, 2010. ISBN: 9783943063042

[RIED94] RIEDMILLER, M.: *Advanced supervised learning in multi-layer perceptrons — From backpropagation to adaptive learning algorithms.* In: Computer Standards & Interfaces, Vol. 16, 1994, No. 3, pp. 265–278. DOI: 10.1016/0920-5489(94)90017-5

[RODR14] RODRÍGUEZ RIPOLL, M.; HEINDL, F.; LECHNER, C.; HABERSOHN, C.; JECH, M.; BLEICHER, F.: *Improving wear resistance of functional surfaces using the machine hammer peening technique.* In: Tribology - Materials, Surfaces & Interfaces, Vol. 8, 2014, No. 1, pp. 21–26. DOI: 10.1179/1751584X14Y.0000000063

[ROTH12] ROTH, I.: *Untersuchungen zum Ausbreitungsverhalten mikrostrukturell kurzer Ermüdungsrisse in metastabilem austenitischen Edelstahl.* PhD-Thesis. Universität Siegen, 2012. URI: https://dspace.ub.uni-siegen.de/handle/ubsi/612

[RUGE07] RUGE, J.; WOHLFAHRT, H.: *Technologie der Werkstoffe. Herstellung-Verarbeitung-Einsatz*. Vol. 8, Wiesbaden: Friedr. Vieweg & Sohn Verlag, 2007. ISBN: 9783834802866

[SAND20] SANDEEP, C. S.; SENETAKIS, K.; CHEUNG, D. K. H.; CHOI, C. E.; WANG, Y.; COOP, M.; NG, C.W.W.: *Experimental study on the coefficient of restitution of grain against block interfaces for natural and engineered materials*. In: Canadian Geotechnical Journal, 2020, pp. 1–14. DOI: 10.1139/cgj-2018-0712

[SAND69] SANDERSON, G. P.; LLEWELLYN, D. T.: *Mechanical Properties of Standard Austenitic Stainless Steels in the Temperature Range -196 to +800 °C*. In: Journal of the Iron and Steel Institute, Vol. 207, 1969, No. 8, pp. 1129–1140

[SANT06] SANTACREU, P.-O.; GLEZ, J.-C.; CHINOUILH, G.; FRÖHLICH, T.: *Behaviour Model of Austenitic Stainless Steels for Automotive Structural Parts*. In: Steel Research International, Vol. 77, 2006, No. 9-10, pp. 686–691. DOI: 10.1002/srin.200606448

[SCHA11] SCHAJER, G. S.; RICKERT, T. J.: *Incremental Computation Technique for Residual Stress Calculations Using the Integral Method*. In: Experimental Mechanics, Vol. 51, 2011, No. 7, pp. 185–191. DOI: 10.1007/s11340-010-9408-5

[SCHA49] SCHAEFFLER, A. L.: *Constitution diagram for stainless steel weld metal*. In: Metal Progress, Vol. 56, 1949, No. 11, 680-680B

[SCHE13] SCHEIL, J.; MÜLLER, C.; STEITZ, M.; GROCHE, P.: *Influence of Process Parameters on Surface Hardening in Hammer Peening and Deep Rolling*. In: Key Engineering Materials, Vol. 554-557, 2013, pp. 1819–1827. DOI: 10.4028/www.scientific.net/KEM.554-557.1819

[SCHE16] SCHEIL, J.: *Entwicklung von Austenitisch-Ferritischem Gusseisen (ADI) aus EN-JS2070: Mikrostruktur, mechanische Eigenschaften und deren Auswirkung auf die Oberflächenbearbeitung durch das Maschinelle Oberflächenhämmern*. PhD-Thesis. TU Darmstadt, 2016. URN: nbn:de:tuda-tuprints-54232

[SCHO01] SCHOß; V.: *Martensitische Umwandlung und Ermüdung austenitischer Edelstähle, Gefügeveränderungen und Möglichkeiten der Früherkennung von Ermüdungsschädigungen*. PhD-Thesis. TU Bergakademie Freiberg, 2001. URL: https://nbn-resolving.org/urn:nbn:de:swb:105-9594166

[SCHO86] SCHOLTES, B.; MACHERAUCH, E.: *Auswirkungen mechanischer Randschichtverformungen auf das Festigkeitsverhalten metallischer Werkstoffe*. In: Zeitschrift für Metallkunde, Vol. 77, 1986, No. 5, pp. 322–337

[SCHO91] SCHOLTES, B.: *Eigenspannungen in mechanisch randschichtverformten Werkstoffzuständen - Ursachen, Ermittlung und Bewertung,* 1991. ISBN: 9783883551708

[SCHO93] SCHOLTES, B.; VÖHRINGER, O.: *Ursachen, Ermittlung und Bewertung von Randschichtveränderungen durch Kugelstrahlen*. In: Materialwissenschaft und Werkstofftechnik, Vol. 24, 1993, No. 12, pp. 421–431. DOI: 10.1002/mawe.19930241206

[SCHR75] SCHRAMM, R. E.; REED, R. P.: *Stacking fault energies of seven commercial austenitic stainless steels*. In: Metallurgical and Materials Transactions A, Vol. 6, 1975, No. 7, pp. 1345–1351. DOI: 10.1007/BF02641927

[SCHU06] SCHULZE, V.; SCHWING, J. K.: *Modern mechanical surface treatment. States, stability, effects,* Weinheim: Wiley-VCH, 2006. ISBN: 9783527313716

[SCHU09] SCHUMANN, H.; OETTEL, H.: *Metallografie*. Vol. 14, Weinheim: Wiley-VCH, 2009. ISBN: 9783527306794

[SCHU16] SCHULZE, V.; BLEICHER, F.; GROCHE, P.; GUO, Y. B.; PYUN, Y. S.: *Surface modification by machine hammer peening and burnishing*. In: CIRP Annals, Vol. 65, 2016, No. 2, pp. 809–832. DOI: 10.1016/j.cirp.2016.05.005

[SCHU75] SCHUMANN, H.: *Verformungsinduzierte Martensitbildung in metastabilen austenitischen Stählen*. In: Kristall und Technik, Vol. 10, 1975, No. 4, pp. 401–411. DOI: 10.1002/crat.19750100409

[SHRI95] SHRINIVAS, V.; VARMA, S. K.; MURR, L. E.: *Deformation-induced martensitic characteristics in 304 and 316 stainless steels during room-temperature rolling*. In: Metallurgical and Materials Transactions A, Vol. 26, 1995, No. 3, pp. 661–671. DOI: 10.1007/BF02663916

[SIEB17] SIEBERTZ, K.; VAN BEBBER, D.; HOCHKIRCHEN, T.: *Statistische Versuchsplanung. Design of Experiments (DoE)*. Vol. 2, Berlin, Heidelberg: Vieweg, 2017. ISBN: 9783662557433

[SILB20] SILBITZ GROUP: *Windkrafttechnik im Überblick,* 2020. URL: https://www.silbitz-group.com/de/windkrafttechnik.html

[SING17] SINGH, L.; KHAN, R. A.; AGGARWAL, M. L.: *Influence of Residual Stress on Fatigue Design of AISI 304 Stainless Steel*. In: The Journal of Engineering Research, Vol. 8, 2017, No. 1, pp. 44–52. DOI: 10.24200/tjer.vol8iss1pp44-52

[SKEL87] SKELTON, R. P.: *High Temperature Fatigue. Properties and Prediction,* Dordrecht (Netherlands): Springer, 1987. ISBN: 9789401080460

[SMID09] SMIDA, M.; SENK, B.; STRÄHUBER, V.: *Versuche und Anwendung hochfrequenter Hämmerverfahren als Schweißnahtnachbehandlung*. In: Stahlbau, Vol. 78, 2009, No. 9, pp. 650–661. DOI: 10.1002/stab.200910080

[SOND90] SONDERGAARD, R.; CHANEY, K.; BRENNEN, C. E.: *Measurements of Solid Spheres Bouncing Off Flat Plates*. In: Journal of Applied Mechanics, Vol. 57, 1990, No. 3, pp. 694–699. DOI: 10.1115/1.2897079

[SPEN09] SPENCER, K.; VÉRON, M.; YU-ZHANG, K.; EMBURY, J. D.: *The strain induced martensite transformation in austenitic stainless steels: Part 1 – Influence of temperature and strain history*. In: Materials Science and Technology, Vol. 25, 2009, No. 1, pp. 7–17. DOI: 10.1179/174328408X293603

[SPRI06] SPRINGUB, B.: *Semi-analytische Betrachtung des Tiefziehens rotationssymmetrischer Bauteile unter Berücksichtigung der Martensitevolution*. PhD-Thesis. Universität Hannover, 2006. ISBN: 9783939026198

[STEI13] STEITZ, M.; SCHEIL, J.; MÜLLER, C.; GROCHE, P.: *Effect of Process Parameters on Surface Roughness in Hammer Peening and Deep Rolling*. In: Key Engineering Materials, Vol. 554-557, 2013, pp. 1887–1901. DOI: 10.4028/www.scientific.net/KEM.554-557.1887

[STEI16] STEITZ, M.: *Tribologisch günstige Oberflächenstrukturierung von Tiefziehwerkzeugen mittels maschinellem Oberflächenhämmern*. PhD-Thesis. TU Darmstadt, 2016. ISBN: 9783844046915

[STIC16] STICHT, P.; STEITZ, M.; GROCHE, P.: *Prediction and Experimental Validation of an Impact Energy Threshold for Mechanical Surface Smoothing*. In: Procedia CIRP, Vol. 45, 2016, pp. 159–162. DOI: 10.1016/j.procir.2016.02.353

[STRO00] STRONGE, W. J.: *Impact mechanics,* Cambridge: Cambridge University Press, 2000. ISBN: 9780511626432

[SUND19] SUNDAR, R.; GANESH, P.; RAM KISHOR, G.; RAGVENDRA, G.; PANT, B. K.; VIVEKANAND, K.; RANGANATHAN, K.; RAKESH, K.; BINDRA, K. S.: *Laser Shock Peening and its Applications: A Review*. In: Lasers in Manufacturing and Materials Processing, Vol. 6, 2019, No. 4, pp. 424–463. DOI: 10.1007/s40516-019-00098-8

[SUNG10] SUNG, J. H.; KIM, J. H.; WAGONER, R. H.: *A plastic constitutive equation incorporating strain, strain-rate, and temperature*. In: International Journal of Plasticity, Vol. 26, 2010, No. 12, pp. 1746–1771. DOI: 10.1016/j.ijplas.2010.02.005

[SURE04] SURESH, S.: *Fatigue of materials*. Vol. 2, Cambridge: Cambridge Univ. Press, 2004. ISBN: 9780521578479

[SUZU76] SUZUKI, T.; KOJIMA, H.; SUZUKI, K.; HASHIMOTO, T.; KOIKE, S.; ICHIHARA, M.: *Plastic deformation and martensitic transformation in an iron-base alloy*. In: Scripta Metallurgica, Vol. 10, 1976, No. 4, pp. 353–358. DOI: 10.1016/0036-9748(76)90091-0

[TABO00] TABOR, D.: *The Hardness of Metals,* Oxford: Oxford University Press, 2000. ISBN: 9780198507765

[TAKE01] TAKEUCHI, S.: *The mechanism of the inverse Hall-Petch relation of nanocrystals*. In: Scripta Materialia, Vol. 44, 2001, No. 8-9, pp. 1483–1487. DOI: 10.1016/S1359-6462(01)00713-8

[TALO07] TALONEN, J.: *Effect of Strain-Induced α'-Martensite Transformation on Mechanical Properties of Metastable Austenitic Stainless Steels*. PhD-Thesis. Helsinki University of Technology, 2007. ISBN: 9789512287796

[TAYL34] TAYLOR, G. I.; QUINNEY, H.: *The latent energy remaining in a metal after cold working*. In: Proceedings of the Royal Society of London A, Vol. 143, 1934, No. 849, pp. 307–326. DOI: 10.1098/rspa.1934.0004

[TJON04] TJONG, S. C.; CHEN, H.: *Nanocrystalline materials and coatings*. In: Materials Science and Engineering: R: Reports, Vol. 45, 2004, No. 1-2, pp. 1–88. DOI: 10.1016/j.mser.2004.07.001

[TOMI95] TOMITA, Y.; IWAMOTO, T.: *Constitutive modeling of trip steel and its application to the improvement of mechanical properties*. In: International Journal of Mechanical Sciences, Vol. 37, 1995, No. 12, pp. 1295–1305. DOI: 10.1016/0020-7403(95)00039-Z

[TOTT02] TOTTEN, G. E.; HOWES, M. A. H.; INOUE, T.: *Handbook of residual stress and deformation of steel,* Materials Park (Ohio, USA): ASM International, 2002. ISBN: 9780871707291

[TRAU13] TRAUTH, D.; KLOCKE, F.; SCHONGEN, F.; SHIROBOKOV, A.: *Analyse und Modellierung der Schlagkraft beim elektro-dynamischen Festklopfen zur kraftbasierten Prozessauslegung*. In: UTF Science, Vol. III, 2013, pp. 1–8

[TRAU15A] TRAUTH, D.; KLOCKE, F.; TERHORST, M.; MATTFELD, P.: *Physicochemical Analysis of Machine Hammer Peened Surface Structures for Deep Drawing: Determination of the Work of Adhesion and Spreading Pressure of Lubrication to Surface Structure*. In: Journal of Tribology, Vol. 137, 2015, No. 2, pp. 1–7. DOI: 10.1115/1.4029199

[TRAU15B] TRAUTH, D.; FEUERHACK, A.; MATTFELD, P.; KLOCKE, F.: *Analysis of the Velocity Distribution of an Elliptic Surface Structure Manufactured by Machine Hammer Peening*. In: Tribology Letters, Vol. 60, 2015, No. 19, pp. 1–13. DOI: 10.1007/s11249-015-0595-1

[TRAU16A] TRAUTH, D.: *Tribology of Machine Hammer Peened Tool Surfaces for Deep Drawing*. PhD-Thesis. RWTH Aachen University, 2016. ISBN: 9783863594244

[TRAU16B] TRAUTH, D.; KLOCKE, F.; WELLING, D.; TERHORST, M.; MATTFELD, P.; KLINK, A.: *Investigation of the surface integrity and fatigue strength of Inconel718 after wire EDM and machine hammer peening*. In: International Journal of Material Forming, Vol. 9, 2016, No. 5, pp. 635–651. DOI: 10.1007/s12289-015-1249-4

[TSUC00] TSUCHIDA, N.; TOMOTA, Y.: *A Micromechanic Modeling for Transformation Induced Plasticity in Steels*. In: Materials Science and Engineering: A, Vol. 285, 2000, No. 1, pp. 345–352. DOI: 10.1016/S0921-5093(00)00688-2

[TWAR13] TWARDOWSKI, R.: *Mikrostrukturelle Beschreibung von Verformung und Schädigung hochmanganhaltiger Stähle mit TRIP- und TWIP-Effekt*. PhD-Thesis. RWTH Aachen University, 2013

[USDE19] U.S. ENERGY INFORMATION ADMINISTRATION: *International Energy Outlook 2019 with Projections to 2050.* U.S. Department of Energy, Washington D.C. (USA), 2019

[VAND81] VANDE WALLE, L. J.: *Residual Stress for Designers and Metallurgists. Proceedings,* Metals Park (Ohio, USA): The American Society for Metals, 1981. ISBN: 9780871701060

[VARM94] VARMA, S. K.; KALYANAM, J.; MURK, L. E.; SRINIVAS, V.: *Effect of grain size on deformation-induced martensite formation in 304 and 316 stainless steels during room temperature tensile testing*. In: Journal of Materials Science Letters, Vol. 13, 1994, No. 2, pp. 107–111. DOI: 10.1007/BF00416816

[VDI18] VEREIN DEUTSCHER INGENIEURE (VDI): *VDI 3416: Maschinelles Oberflächenhämmern - Grundlagen.* Februar 2018

[VÖHR77] VÖHRINGER, O.; MACHERAUCH, E.: *Struktur und mechanische Eigenschaften von Martensit*. In: HTM Härtereitechnik, Vol. 32, 1977, pp. 153–168

[VÖHR87] VÖHRINGER, O.: *Changes in the state of the material by shot peening*. In: Wohlfahrt, H.; Kopp, R.; Vöhringer, O. (Ed.): Shot Peening.: Science, Technology, Application. Oberursel: DGM-Informationsgesellschaft Verlag, 1987, pp. 185–204

[WAGN94] WAGNER, L.; GREGORY, J. K.: *Thermomechanical Surface Treatment of Titanium Alloys*. In: Materials Science Forum, Vol. 163-165, 1994, pp. 159–172. DOI: 10.4028/www.scientific.net/MSF.163-165.159

[WANG16] WANG, G. Q.; LEI, M. K.; GUO, D. M.: *Interactions between Surface Integrity Parameters on AISI 304 Austenitic Stainless Steel Components by Ultrasonic Impact Treatment*. In: Procedia CIRP, Vol. 45, 2016, pp. 323–326. DOI: 10.1016/j.procir.2016.02.351

[WANG20A] WANG, C.; WANG, C.; WANG, L.; LAI, Y.; LI, K.; ZHOU, Y.: *A dislocation density–based comparative study of grain refinement, residual stresses, and surface roughness induced by shot peening and surface mechanical attrition treatment*. In: The International Journal of Advanced Manufacturing Technology, Vol. 108, 2020, No. 1-2, pp. 505–525. DOI: 10.1007/s00170-020-05413-8

[WANG20B] WANG, J. L.; HUANG, M. H.; XI, X. H.; WANG, C. C.; XU, W.: *Characteristics of nucleation and transformation sequence in deformation-induced martensitic transformation*. In: Materials Characterization, Vol. 163, 2020, p. 110234–110234. DOI: 10.1016/j.matchar.2020.110234

[WARE13] WARENZEICHENVERBAND EDELSTAHL ROSTFREI: *Edelstahl Rostfrei in der Umwelttechnik,* 2013. URL: https://www.wzv-rostfrei.de/presse/detail/edelstahl-rostfrei-in-der-umwelttechnik

[WEBE12] WEBER, S.; MARTIN, M.; THEISEN, W.: *Impact of heat treatment on the mechanical properties of AISI 304L austenitic stainless steel in high-pressure hydrogen gas*. In: Journal of Materials Science, Vol. 47, 2012, No. 16, pp. 6095–6107. DOI: 10.1007/s10853-012-6526-8

[WEBS66] WEBSTER, G. A.: *A widely applicable dislocation model of creep*. In: The Philosophical Magazine: A Journal of Theoretical Experimental and Applied Physics, Vol. 14, 1966, No. 130, pp. 775–783. DOI: 10.1080/14786436608211971

[WEIL19] WEILER, T.; KRALL, S.; BLEICHER, F.: *Linear reluctance actuator and tools/drives designed to comprise same. International Patent PCT/EP2018/081206,* 2019

[WEIS94] WEISS, A.: *Die Martensitbildung in metastabilen austenitischen Stählen unter der Wirkung äusserer Spannungen und ihr Einfluss auf das Verfestigungs- und Fließverhalten. Abschlussbericht zum Forschungsvorhaben AIF 233 D/I,* Freiberg, 1994

[WESS17] WESSELAK, V.; SCHABBACH, T.; LINK, T.; FISCHER, J.: *Handbuch Regenerative Energietechnik*. Vol. 3, Berlin, Heidelberg: Springer, 2017. ISBN: 9783662530733

[WHIT28] WHITTAKER, J. M.: *The "Fourier" Theory of the Cardinal Function*. In: Proceedings of the Edinburgh Mathematical Society, Vol. 1, 1928, No. 3, pp. 169–176. DOI: 10.1017/S0013091500013511

[WIED10] WIED, J.: *Cold forging device for surface of metal workpiece, has hollow piston element with pneumatic hammer adjustably arranged in adapter housing to position pneumatic hammer with impact head during pre-loading on metal surface. German Patent DE102010019547,* 2010

[WIED11A] WIED, J.: *Oberflächenbehandlung von Umformwerkzeugen durch Festklopfen*. PhD-Thesis. TU Darmstadt, 2011. URI: https://tuprints.ulb.tu-darmstadt.de/id/eprint/2487

[WIED11B] WIED, J.; SCHEIL, J.; KLAMSER, M.; BERGER, C.: *Impact experiments and finite element simulation of surface roughness reduction by machine hammer peening*. In: Materialwissenschaft und Werkstofftechnik, Vol. 42, 2011, No. 9, pp. 827–832. DOI: 10.1002/mawe.201100768

[WOHL00] WOHLFAHRT, H.: *Mechanische Oberflächenbehandlungen. Grundlagen, Bauteileigenschaften, Anwendungen,* Weinheim: Wiley-VCH, 2000. ISBN: 9783527301186

[WÖHL58] WÖHLER, A.: *Bericht über die Versuche, welche auf der Königl. Niederschlesisch-Märkischen Eisenbahn mit Apparaturen zum Messen der Biegung und Verdrehung von Eisenbahnwagen-Achsen während der Fahrt angestellt wurden*. In: Zeitschrift für Bauwesen, Vol. 8, 1858, pp. 642–652

[WOHL83A] WOHLFAHRT, H.: *Ein Modell zur Vorhersage kugelstrahlbedingter Eigenspannungszustände. Eigenspannungen: Entstehung-Messung-Bewertung*. In: DGM Informationsgesellschaft, 1983, pp. 301–319

[WOHL83B] MACHERAUCH, E.; HAUK, V.: *Ein Modell zur Vorhersage kugelstrahlbedingter Eigenspannungszustände*. In: Macherauch, E.; Hauk, V. (Ed.): Eigenspannungen - Band 2. Oberursel: Deutsche Gesellschaft für Materialkunde e. V., pp. 301–319

[WORL20] WORLD ECONOMIC FORUM: *The Global Risks Report 2020. Insight Report - 15th Edition.* World Economic Forum, Cologny/Geneva, 2020

[YE14] YE, C.; TELANG, A.; GILL, A. S.; SUSLOV, S.; IDELL, Y.; ZWEIACKER, K.; WIEZOREK, J. M.K.; ZHOU, Z.; QIAN, D.; MANNAVA, S. R.; VASUDEVAN, V. K.: *Gradient nanostructure and residual stresses induced by Ultrasonic Nano-crystal Surface Modification in 304 austenitic stainless steel for high strength and high ductility*. In: Materials Science and Engineering: A, Vol. 613, 2014, pp. 274–288. DOI: 10.1016/j.msea.2014.06.114

[ZACK67] ZACKAY, V. F.; PARKER, E. R.; FAHR, D.; BUSH, R.: *The enhancement of ductility in high-strength steels*. In: Transactions of the American Society for Metals (ASM), Vol. 60, 1967, pp. 252–257

[ZEHE93] ZEHETBAUER, M.: *Cold work hardening in stages IV and V of F.C.C. metals—II. Model fits and physical results*. In: Acta Metallurgica et Materialia, Vol. 41, 1993, No. 2, pp. 589–599. DOI: 10.1016/0956-7151(93)90089-B

Appendix

Anhang

Appendix for chapter 4

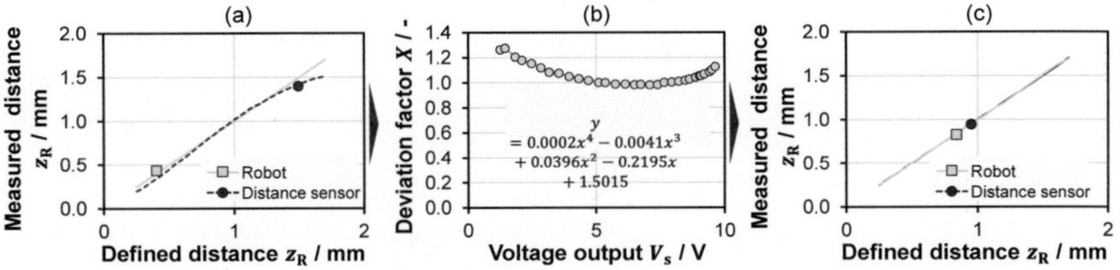

Figure A.1: Deviation between robot and distance sensor (a), deviation factor depending on output voltage (b) and calibrated sensor measurement (c)
Abweichung zwischen Roboter und Abstandssensor (a), Abweichungsfaktor in Abhängigkeit der Ausgangsspannung (b) und kalibrierte Sensormessung (c)

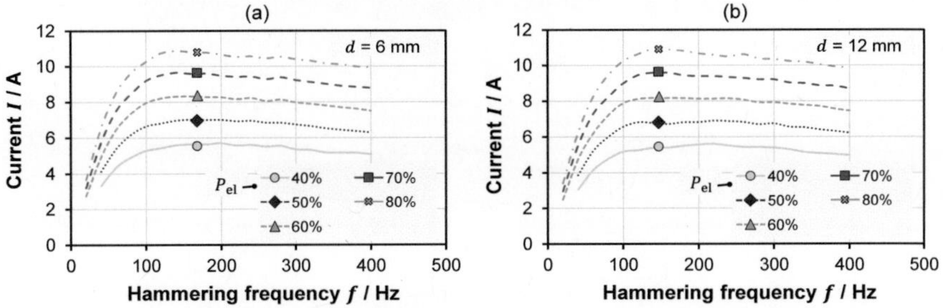

Figure A.2: Analysis of the current intensity for a hammer head diameter d = 6 mm (a) and d = 12 mm (b)
Analyse der Stromstärke für einen Hammerkopfdurchmesser d = 6 mm (a) und d = 12 mm (b)

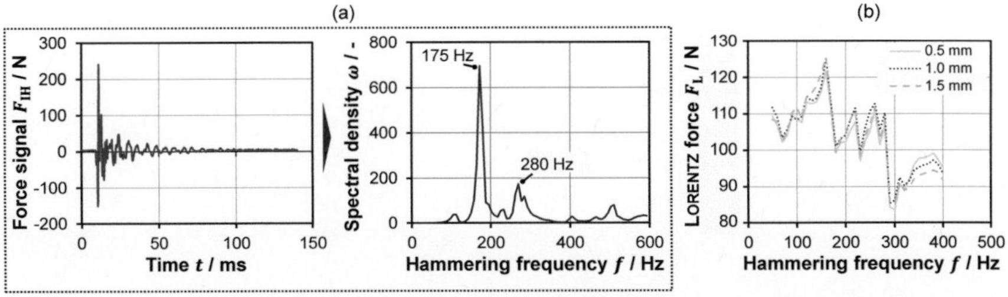

Figure A.3: Analysis of the MHP system natural frequencies (a) and determination of the LORENTZ force (b)
Analyse der MOH-Systemeigenfrequenzen (a) und Bestimmung der LORENTZkraft (b)

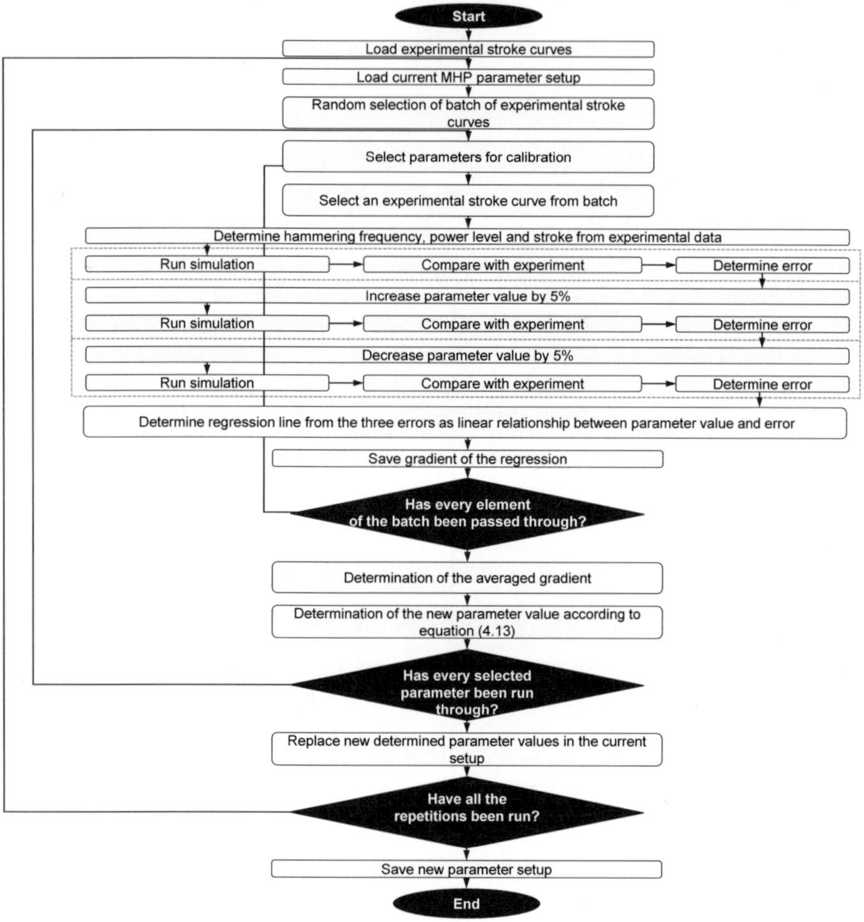

Figure A.4: Flowchart of the learning function according to the gradient method for the MHP kinetics model in MATLAB

Ablaufdiagramm der Anlernfunktion nach dem Gradientenverfahren für das MOH-Kinetikmodell in MATLAB

Appendix for chapter 5

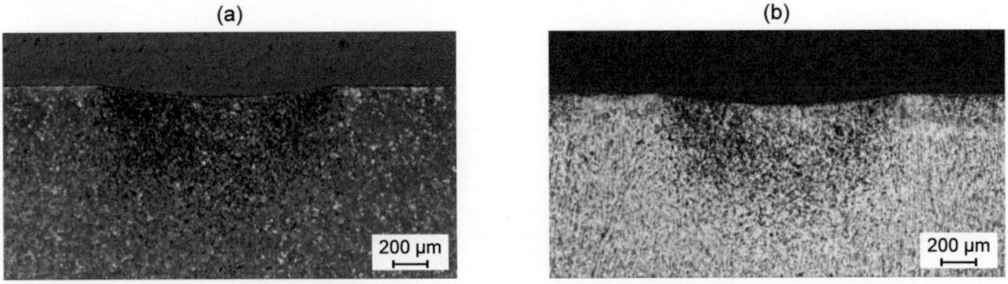

Figure A.5: Etched micrograph with plastically deformed area under indentation (a) and re-polished micrograph (b)
Angeätztes Schliffbild mit plastisch deformierten Bereich unter Eindruck (a) und nachpoliertes Schliffbild (b)

Appendix for chapter 6

Table A.1: Measured and calculated values HV5 in Chapter 6
Gemessene und berechnete HV5-Werte in Kapitel 6

MHP setup no.	Measured HV5	Calculated HV5	Deviation σ_{HV5} / %
1	487.7	487.3	0.09
2	481.4	483.6	-0.46
3	491.4	495.6	-0.86
4	489.4	483.4	1.23
5	402.9	404.8	-0.48
6	403.2	399.5	0.92
7	418.0	412.3	1.37
8	391.0	398.5	-1.92
9	398.7	395.8	0.73
10	389.9	391.1	-0.29
11	406.2	405.3	0.22
12	389.4	392.0	-0.69
13	316.9	318.4	-0.45
14	311.7	312.1	-0.12
15	324.8	327.1	-0.73
16	316.5	312.4	1.32
MP	412.3	401.2	2.69

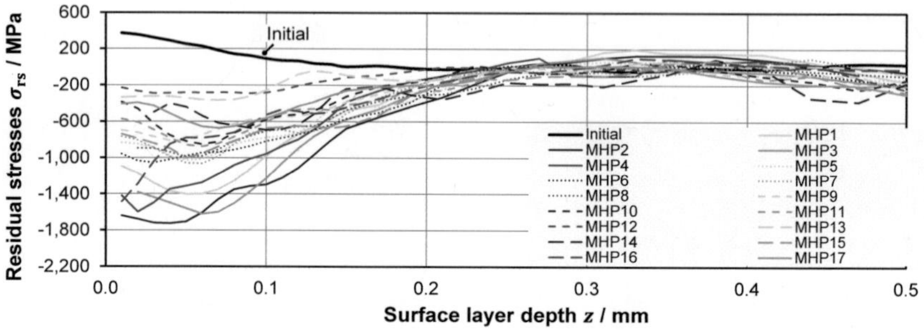

Figure A.6: Overview of all experimental residual stress measurements using the borehole method

Übersicht aller experimenteller Eigenspannungsmessungen mittels Bohrlochmethode

---

User Subroutine VUMAT

```
      Subroutine Vumat (
c Read only -
      1 nblock, ndir, nshr, nstatev, nfieldv, nprops, lanneal,
      2 stepTime, totalTime, dt, cmname, coordMP, charLength,
      3 props, density, strainInc, relSpinInc,
      4 tempOld, stretchOld, defgradOld, fieldOld,
      5 stressOld, stateOld, enerInternOld, enerInelasOld,
      6 tempNew, stretchNew, defgradNew, fieldNew,
c Write only -
      7 stressNew, stateNew, enerInternNew, enerInelasNew )
c
      include 'vaba_param.inc'
c
      dimension props(nprops), density(nblock), coordMP(nblock),
      1 charLength(nblock), strainInc(nblock, ndir+nshr),
      2 resSpinInc(nblock, nshr), tempOld(nblock),
      3 stretchOld(nblock, ndir+nshr), defgradOld(nblock, ndir+nshr+nshr),
      4 fieldOld(nblock, nfieldv), stressOld(nblock,ndir+nshr),
      5 stateOld(nblock,nstatev), enerInternOld(nblock),
      6 enerInelasOld(nblock), tempNew(nblock),
      7 stretchNew(nblock, ndir+nshr), defgradNew(nblock, ndir+nshr+nshr),
      8 fieldNew(nblock, nfieldv), stressNew(nblock, ndir+nshr),
      9 stateNew(nblock, nstatev), enerInternNew(nblock),
      1 enerInelasNew(nblock)
c
      character*8 cmname
c
      parameter (zero = 0.d0, one = 1.d0, two = 2.d0,three = 3.D0,five=5.d0, third = 1.d0 /
      3.d0, half = 0.5d0, op5 = 1.5d0, SIX = 6.D0,M=3.06d0,alpha = 0.1716769d0,
```

```
      1 beta = 0.0020981772d0,gammatilde=2.5d0,burgers=2.56d-10,
      k0=3.6026512365d0,ak = 30.d0, am = 1200.d0,an = 50.d0, zlamda = 100.d0, zkzero
      = 1.2d0, a = 0.33d0, f0=0.33d0,f1=0.06d0,dcell=1.6d-4)
c
c     For plane strain, axisymmetric, and 3D cases using the J2 Mises Plasticity with
c     piecewise-linear isotropic hardening.
c     The state variable is stored as:
c
      state(*,1) = equivalent plastic strain
c
c     User needs to input:
c     props(1) Young's modulus
c     props(2) Poisson's ratio
c     props(3..) syield and hardening data
c     calls vuhard for curve of yield stress vs. plastic strain
c
      e = props(1)
      xnu = props(2)
      syield= props(3)
      two_thirds=two/three
      three_halfs=three*half
      five_halfs=five*half
      burgersinv=one/burgers
      twomu = e / ( one + xnu )
      alamda = xnu * twomu / ( one - two * xnu )
      thremu = op5 * twomu
c     nvalue = nprops/2-1,
      factor1=sqrtwo/THREE
      factor2=ONE/sqrt(THREE)
      sqrthree=sqrt(THREE)
      G=TWOMU/TWO
c
      if (stepTime .eq. zero) then
            do k = 1, nblock
              trace = strainInc(k,1) + strainInc(k,2) + strainInc(k,3)
              stressNew(k,1) = stressOld(k,1)+ twomu * strainInc(k,1) + alamda * trace
              stressNew(k,2) = stressOld(k,2)+ twomu * strainInc(k,2) + alamda * trace
              stressNew(k,3) = stressOld(k,3)+ twomu * strainInc(k,3) + alamda * trace
              stressNew(k,4) = stressOld(k,4) + twomu * strainInc(k,4)
              if ( nshr .gt. 1 ) then
                  stressNew(k,5)=stressOld(k,5) + twomu * strainInc(k,5)
                  stressNew(k,6)=stressOld(k,6) + twomu * strainInc(k,6)
              end if
              stateOld(k,2)=1.0E+12
              stateOld(k,3)=1.0E+12
              stateOld(k,4)=props(3)

            end do
```

```
        else
            do k = 1, nblock
              peeqOld=stateOld(k,1)
              yieldOld = stateold(k,4)
              trace = strainInc(k,1) + strainInc(k,2) + strainInc(k,3)
              s11 = stressOld(k,1) + twomu * strainInc(k,1) + alamda * trace
              s22 = stressOld(k,2) + twomu * strainInc(k,2) + alamda * trace
              s33 = stressOld(k,3) + twomu * strainInc(k,3) + alamda * trace
              s12 = stressOld(k,4) + twomu * strainInc(k,4)
              if ( nshr .gt. 1 ) then
                  s13 = stressOld(k,5) + twomu * strainInc(k,5)
                  s23 = stressOld(k,6) + twomu * strainInc(k,6)
              end if
c
              smean = third * ( s11 + s22 + s33 )
              s11 = s11 - smean
              s22 = s22 - smean
              s33 = s33 - smean
              if ( nshr .eq. 1 ) then
                  vmises = sqrt( op5*(s11*s11+s22*s22+s33*s33+two*s12*s12) )
              else
                  vmises = sqrt( op5 * ( s11 * s11 + s22 * s22 + s33 * s33 + two * s12 *
                  s12 + two * s13 * s13 + two * s23 * s23 ) )
              end if
c
              sigdif = vmises - yieldOld
              facyld = zero
              if ( sigdif .gt. zero ) facyld = one
              deqps = facyld * sigdif / ( thremu + hard )
c
c
              DEQPSDT=deqps/DT
              DGAMMARDT=DEQPSDT*M+1d-15
              f= f0 + f1 * exp(-STATEnew(k,1)*M/five_halfs)
              rhototold=f*STATEOLD(k,3)+(one-f)*STATEOLD(k,2)
              dold=160.0/sqrt(rhototold)
                          rhocr1=alpha*fac-
tor2*sqrt(STATEOLD(k,3))*burgersinv*DGAMMARDT
              rhocr1_1=alpha*factor2
              rhocr1_2=sqrt(STATEOLD(k,3))
              rhocr1_3=burgersinv*DGAMMARDT
              rhocr1=rhocr1_1*rhocr1_2*rhocr1_3
              rhocr2_1=-beta*SIX*DGAMMARDT
              rhocr2_2=burgersinv/dold
              rhocr2_3=((one-f)**THIRD)
              rhocr2=rhocr2_1*rhocr2_2/rhocr2_3
              rhocr3_1=-k0*(DGAMMARDT/zlamda)**(-one/an)
c             rhocr3_1_1=(DGAMMARDT/zlamda)
```

```
c                rhocr3_1_2=(abs(rhocr3_1_1)**(-0.02))
c                rhocr3_1=rhocr3_1_2
                 rhocr3_2=DGAMMARDT*STATEOLD(k,2)
                 rhocr3=rhocr3_1*rhocr3_2

                 rhocr=rhocr1+rhocr2+rhocr3
                 rhowr=sqrthree*beta*DGAMMARDT*(one-
                 f)*sqrt(STATEOLD(k,3))/f*burgersinv+beta*SIX*DGAMMARDT*burgersinv/
                 dold*((one-f)**TWO_THIRDS)/f-k0*(DGAMMARDT/zlamda)**(-
                 one/an)*DGAMMARDT*STATEOLD(k,3)
                 rhocr=rhocr*facyld
                 rhowr=rhowr*facyld

                 STATENEW(k,2) = STATEOLD(k,2) + rhocr*DT
                 STATENEW(k,3) = STATEOLD(k,3) + rhowr*DT
                 Statenew(k,5) = f*STATEOLD(k,3)+(one-f)*STATEOLD(k,2)
                 statenew(k,6) = log(1+statenew(k,1))
                 statenew(k,7) = dold

                 tauw1=a*G*burgers
                 tauw2=sqrt(STATENEW(k,3))
                 tauw3=(DGAMMARDT/zlamda)
                 tauw4=tauw3**(1/am)
                 rauw=tau1*tauw2*tauw4
                         tauc=a*G*bur-
gers*sqrt(STATENEW(k,2))*(DGAMMARDT/zlamda)**(1/am)
                 tau=f*tauw+(1-f)*tauc
                 phi2=tau*M
c
c     ----------Update the stress----------
c
                 yieldNew = phi2+props(3)
                 statenew(k,4)=yieldNew
                 factor = yieldNew / ( yieldNew + thremu * deqps )
                 stressNew(k,1) = s11 * factor + smean
                 stressNew(k,2) = s22 * factor + smean
                 stressNew(k,3) = s33 * factor + smean
                 stressNew(k,4) = s12 * factor
                 if ( nshr .gt. 1 ) then
                    stressNew(k,5) = s13 * factor
                    stressNew(k,6) = s23 * factor
                 end if
c
c     ----------Update the state variables----------
c
                 stateNew(k,1) = stateOld(k,1) + deqps
c
```

```
                rhocr(k)=alpha*sqrt(stateOld(k,3))*peeq(k)*3.00/(sqrt(3.0)*dt*2.55E-10)-
                beta*6*peeq(k)*3.00/((2.55E-10)*dt*stateOld(k,5)*((1-f(k))**(1/3)))-
                zkzero*((peeq(k)*3.00/(100*dt))**(-1/50))*peeq(k)*3.00*stateOld(k,1)/dt
c               rhowr(k)=beta*peeq(k)*3.00*(1-
                f(k))*sqrt(3.0)*sqrt(stateOld(k,2))/(f(k)*dt*2.55E-
                10)+6*beta*peeq(k)/dt*3.00*((1-f(k))**(2/3))/((2.55E-10)*stateOld(k,5)*f(k))-
                zkzero*((peeq(k)*3.00/(100*dt))**(-1/50))*peeq(k)*3.00*stateOld(k,2)/dt

c       ----------Update the specific internal energy----------
c
                if ( nshr .eq. 1 ) then
                  stressPower = half * (( stressOld(k,1) + stressNew(k,1) ) * strainInc(k,1) +
                  ( stressOld(k,2) + stressNew(k,2) ) * strainInc(k,2) +( stressOld(k,3) +
                  stressNew(k,3) ) * strainInc(k,3) ) +( stressOld(k,4) + stressNew(k,4) ) *
                  strainInc(k,4)
                else
                  stressPower = half * (( stressOld(k,1) + stressNew(k,1) ) * strainInc(k,1) +
                  ( stressOld(k,2) + stressNew(k,2) ) * strainInc(k,2) +( stressOld(k,3) +
                  stressNew(k,3) ) * strainInc(k,3) ) +( stressOld(k,4) + stressNew(k,4) ) *
                  strainInc(k,4) +( stressOld(k,5) + stressNew(k,5) ) * strainInc(k,5) +(
                  stressOld(k,6) + stressNew(k,6) ) * strainInc(k,6)
                end if
                enerInternNew(k) = enerInternOld(k) + stressPower / density(k)
c
c       ----------Update the dissipated inelastic specific energy----------
c
                plasticWorkInc = half * ( yieldOld + yieldNew ) * deqps
                enerInelasNew(k) = enerInelasOld(k)+ plasticWorkInc / density(k)
              end do
        end if
c
        return
        end
```

Figure A.7: Program code for user subroutine VUMAT with material model according to LEMIALE ET AL. [LEMI10]
Programm-Code für User-Subroutine VUMAT mit Materialmodell nach LEMIALE ET AL.